SRA
Connecting Math Concepts

Level C Workbook 1

COMPREHENSIVE EDITION

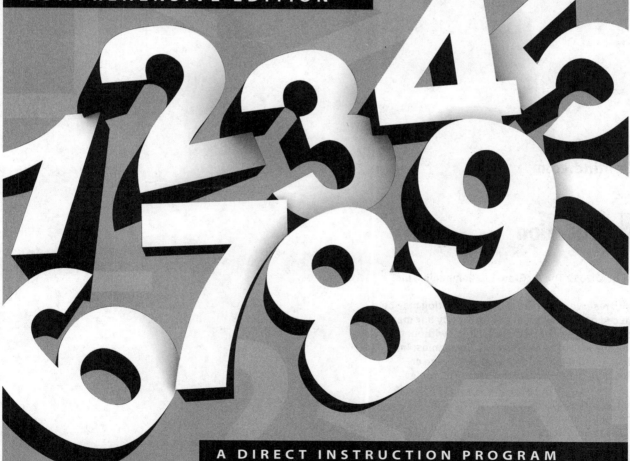

A DIRECT INSTRUCTION PROGRAM

McGraw Hill Education

Bothell, WA • Chicago, IL • Columbus, OH • New York, NY

MHEonline.com

 **Education**

Send all inquiries to:
McGraw-Hill Education
4400 Easton Commons
Columbus, OH 43219

ISBN: 978-0-02-103576-2
MHID: 0-02-103576-8

Printed in the United States of America.

9 QTN 16

The *McGraw-Hill* Companies

Lesson

Name ___Esmeralda___

Part 1

a. $7 + 1 = 8$

b. $7 + 2 = 9$

c. $4 + 1 = 5$

d. $4 + 2 = 6$

e. $9 + 1 = 10$

f. $9 + 2 = 11$

g. $5 + 1 = 6$

h. $5 + 2 = 7$

i. $8 + 1 = 9$

j. $8 + 2 = 10$

Part 2

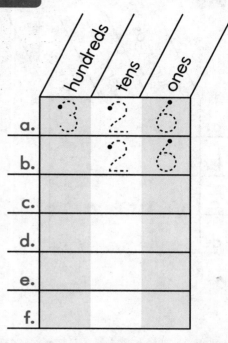

	hundreds	tens	ones
a.	3	2	6
b.		2	6
c.			
d.			
e.			
f.			

Part 3

a. $9 + 0 = 9$

$9 + 1 = 10$

$9 + 2 = 11$

b. $17 + 0 = 17$

$17 + 1 = 18$

$17 + 2 = 19$

c. $34 + 0 = 34$

$34 + 1 = 35$

$34 + 2 = 36$

d. $90 + 0 = 90$

$90 + 1 = 100$

$90 + 2 = 101$

Lesson 2

Name _____

Part 1

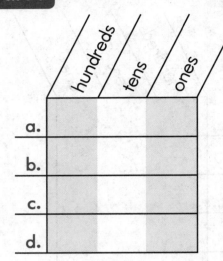

	hundreds	tens	ones
a.			
b.			
c.			
d.			

Part 2

a. $3 + 1 = 4$

b. $3 + 2 = 5$

c. $6 + 1 = 7$

d. $6 + 2 = 8$

e. $4 + 1 = 5$

f. $4 + 2 = 6$

g. $9 + 1 = 10$

h. $9 + 2 = 11$

i. $7 + 1 = 8$

j. $7 + 2 = 9$

Part 3

a. $3 + 0 = \underline{3}$

$3 + 1 = \underline{4}$

$3 + 2 = \underline{5}$

b. $15 + 0 = \underline{15}$

$15 + 1 = \underline{16}$

$15 + 2 = \underline{17}$

c. $88 + 0 = \underline{88}$

$88 + 1 = \underline{89}$

$88 + 2 = \underline{90}$

d. $12 + 0 = \underline{12}$

$12 + 1 = \underline{13}$

$12 + 2 = \underline{14}$

Connecting Math Concepts

Lesson 3

Name _____

Part 1

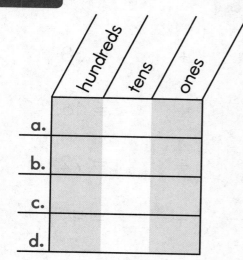

Part 2

a. 7 + 1 = _____

b. 7 + 2 = _____

c. 5 + 1 = _____

d. 5 + 2 = _____

e. 10 + 1 = _____

f. 10 + 2 = _____

g. 3 + 1 = _____

h. 3 + 2 = _____

Part 3

a. ⬜ + ⬜ = 51

b. ⬜ + ⬜ = 15

c. ⬜ + ⬜ = 88

d. ⬜ + ⬜ = 18

e. ⬜ + ⬜ = 81

Part 4

a. 10 + 1 = _____

10 + 2 = _____

10 + 3 = _____

b. 8 + 1 = _____

8 + 2 = _____

8 + 3 = _____

c. 9 + 1 = _____

9 + 2 = _____

9 + 3 = _____

Lesson 4

Name _____

Part 1

a. 8 − 1 = _____

b. 8 − 2 = _____

c. 3 − 1 = _____

d. 3 − 2 = _____

e. 7 − 1 = _____

f. 7 − 2 = _____

g. 9 − 1 = _____

h. 9 − 2 = _____

i. 5 − 1 = _____

j. 5 − 2 = _____

k. 10 − 1 = _____

l. 10 − 2 = _____

Part 2

a. [   ] + [   ] = 38

b. [   ] + [   ] = 16

c. [   ] + [   ] = 72

d. [   ] + [   ] = 49

e. [   ] + [   ] = 15

Part 3

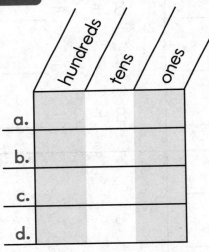

Part 4

a. 2 7
  + 2 1

b. 1 5
  + 6 0

c. 7 3
  + 1 2

d. 8 0
  + 1 8

Part 5

a. 8 + 1 = _____

8 + 2 = _____

8 + 3 = _____

b. 6 + 1 = _____

6 + 2 = _____

6 + 3 = _____

c. 0 + 1 = _____

0 + 2 = _____

0 + 3 = _____

d. 9 + 1 = _____

9 + 2 = _____

9 + 3 = _____

Connecting Math Concepts

Lesson

Name _____

Part 1

a. $6 - 1 =$ _____

b. $6 - 2 =$ _____

c. $4 - 1 =$ _____

d. $4 - 2 =$ _____

e. $10 - 1 =$ _____

f. $10 - 2 =$ _____

g. $8 - 1 =$ _____

h. $8 - 2 =$ _____

i. $9 - 1 =$ _____

j. $9 - 2 =$ _____

k. $12 - 1 =$ _____

l. $12 - 2 =$ _____

Part 2

a. $\begin{array}{r} 1\ 1 \\ +\ 4\ 8 \\ \hline \end{array}$

b. $\begin{array}{r} 2\ 5 \\ +\ 1\ 2 \\ \hline \end{array}$

c. $\begin{array}{r} 7\ 8 \\ +\ 2\ 0 \\ \hline \end{array}$

d. $\begin{array}{r} 1\ 6 \\ +\ 3\ 1 \\ \hline \end{array}$

Part 3

a. ⬜ + ⬜ = 12

b. ⬜ + ⬜ = 93

c. ⬜ + ⬜ = 17

d. ⬜ + ⬜ = 54

Part 4

a. $7 + 1 =$ _____

$7 + 2 =$ _____

$7 + 3 =$ _____

b. $4 + 1 =$ _____

$4 + 2 =$ _____

$4 + 3 =$ _____

c. $9 + 1 =$ _____

$9 + 2 =$ _____

$9 + 3 =$ _____

d. $2 + 1 =$ _____

$2 + 2 =$ _____

$2 + 3 =$ _____

Lesson 6

Name _____

Part 1

a. $10 - 1 = $ _____

b. $10 - 2 = $ _____

c. $3 - 1 = $ _____

d. $3 - 2 = $ _____

e. $11 - 1 = $ _____

f. $11 - 2 = $ _____

g. $6 - 1 = $ _____

h. $6 - 2 = $ _____

i. $12 - 1 = $ _____

j. $12 - 2 = $ _____

k. $7 - 1 = $ _____

l. $7 - 2 = $ _____

Part 2

a.
$$\begin{array}{r} 5\,5 \\ +\,3\,2 \\ \hline \end{array}$$

b.
$$\begin{array}{r} 6\,6 \\ +\,3\,2 \\ \hline \end{array}$$

c.
$$\begin{array}{r} 1\,1 \\ +\,3\,3 \\ \hline \end{array}$$

d.
$$\begin{array}{r} 5\,6 \\ +\,2\,2 \\ \hline \end{array}$$

Part 3

a. ____ + ____ = 53

b. ____ + ____ = 79

c. ____ + ____ = 43

d. ____ + ____ = 91

e. ____ + ____ = 18

Part 4

a.
$$\begin{array}{r} 3 \\ -\,0 \\ \hline \end{array}$$

b.
$$\begin{array}{r} 10 \\ -\,10 \\ \hline \end{array}$$

c.
$$\begin{array}{r} 7 \\ -\,0 \\ \hline \end{array}$$

d.
$$\begin{array}{r} 34 \\ -\,0 \\ \hline \end{array}$$

e.
$$\begin{array}{r} 9 \\ -\,9 \\ \hline \end{array}$$

Connecting Math Concepts

Lesson 6

Name _____

Part 5

a.   4
   + 3

b.   6
   + 3

c.   5
   + 3

d.   8
   + 2

e.   7
   + 3

f.   9
   + 1

g.   5
   + 1

h.   2
   + 3

i.   7
   + 0

j.   6
   + 2

k.   9
   + 3

l.   4
   + 2

Part 6

a.   4 ——— 2 ——→ 6

_____

_____

_____

_____

Lesson 7

Name _____

a. $9 - 2 =$ _____

b. $7 + 2 =$ _____

c. $10 - 2 =$ _____

d. $5 - 2 =$ _____

e. $9 + 2 =$ _____

f. $6 - 2 =$ _____

g. $8 + 2 =$ _____

h. $10 + 2 =$ _____

i. $3 - 2 =$ _____

j. $8 - 2 =$ _____

Part 2

a. $\begin{array}{r} 7 \\ + 3 \\ \hline \end{array}$
b. $\begin{array}{r} 5 \\ + 3 \\ \hline \end{array}$
c. $\begin{array}{r} 8 \\ + 1 \\ \hline \end{array}$
d. $\begin{array}{r} 9 \\ + 0 \\ \hline \end{array}$
e. $\begin{array}{r} 6 \\ + 2 \\ \hline \end{array}$
f. $\begin{array}{r} 4 \\ + 3 \\ \hline \end{array}$

g. $\begin{array}{r} 9 \\ + 2 \\ \hline \end{array}$
h. $\begin{array}{r} 3 \\ + 3 \\ \hline \end{array}$
i. $\begin{array}{r} 6 \\ + 3 \\ \hline \end{array}$
j. $\begin{array}{r} 4 \\ + 1 \\ \hline \end{array}$
k. $\begin{array}{r} 9 \\ + 3 \\ \hline \end{array}$
l. $\begin{array}{r} 5 \\ + 2 \\ \hline \end{array}$

Part 3

a. ▢ + ▢ = 45

b. ▢ + ▢ = 17

c. ▢ + ▢ = 83

d. ▢ + ▢ = 13

e. ▢ + ▢ = 68

Part 4

a. $\begin{array}{r} 44 \\ + 32 \\ \hline \end{array}$
b. $\begin{array}{r} 57 \\ + 21 \\ \hline \end{array}$

c. $\begin{array}{r} 65 \\ + 12 \\ \hline \end{array}$
d. $\begin{array}{r} 33 \\ + 32 \\ \hline \end{array}$

Connecting Math Concepts

Lesson 7

Name _____

Part 5

a. $\quad \underset{\longrightarrow}{6 \qquad 3} 9$

_____

_____

_____

_____

Independent Work

Part 6

a. 407	b. 37	c. 236	d. 910
hundreds digit _____	tens digit _____	hundreds digit _____	tens digit _____
tens digit _____	ones digit _____	ones digit _____	ones digit _____

Part 7

___ 10 ___ ___ ___ 7 ___ ___ ___ ___ ___ 2 ___

Part 8

a. $\begin{array}{r} 5 \\ + 1 \\ \hline \end{array}$ b. $\begin{array}{r} 8 \\ - 8 \\ \hline \end{array}$ c. $\begin{array}{r} 4 \\ - 1 \\ \hline \end{array}$ d. $\begin{array}{r} 7 \\ + 0 \\ \hline \end{array}$ e. $\begin{array}{r} 10 \\ + 1 \\ \hline \end{array}$ f. $\begin{array}{r} 3 \\ - 0 \\ \hline \end{array}$

g. $\begin{array}{r} 6 \\ - 1 \\ \hline \end{array}$ h. $\begin{array}{r} 10 \\ - 0 \\ \hline \end{array}$ i. $\begin{array}{r} 3 \\ - 3 \\ \hline \end{array}$ j. $\begin{array}{r} 12 \\ + 0 \\ \hline \end{array}$ k. $\begin{array}{r} 7 \\ + 1 \\ \hline \end{array}$ l. $\begin{array}{r} 10 \\ - 1 \\ \hline \end{array}$

Lesson 8

Name _____

Part 1

a. 10 – 2 = _____

b. 4 + 2 = _____

c. 9 + 2 = _____

d. 7 – 2 = _____

e. 3 + 2 = _____

f. 8 + 2 = _____

g. 5 – 2 = _____

h. 4 – 2 = _____

i. 6 – 2 = _____

j. 5 + 2 = _____

Part 2

a. 4 + 10 = _____

b. 8 + 10 = _____

c. 2 + 10 = _____

d. 5 + 10 = _____

e. 9 + 10 = _____

f. 6 + 10 = _____

g. 1 + 10 = _____

h. 7 + 10 = _____

i. 3 + 10 = _____

Part 3

a. 4
  + 3
  ———

b. 9
  + 1
  ———

c. 5
  + 2
  ———

d. 2
  + 3
  ———

e. 8
  + 3
  ———

f. 7
  + 0
  ———

g. 6
  + 2
  ———

h. 4
  + 3
  ———

i. 7
  + 3
  ———

j. 8
  + 1
  ———

k. 9
  + 2
  ———

l. 5
  + 3
  ———

Part 4

a. 6 4
  + 2 3
  ———

b. 2 8
  + 3 0
  ———

c. 5 1
  + 3 7
  ———

d. 4 3
  + 1 3
  ———

Lesson 8

Name _____

Part 5 Write 4 facts for each family.

a. $\dfrac{8 \qquad 1}{\longrightarrow} 9$

b. $\dfrac{8 \qquad 2}{\longrightarrow} 10$

_____   _____

_____   _____

_____   _____

Part 6

a.	b.	c.	d.	e.	f.
12	4	9	6	3	5
− 12	+ 1	− 0	− 1	− 0	+ 1

g.	h.	i.	j.	k.	l.
5	10	3	6	12	4
− 1	− 10	− 1	− 0	+ 1	− 4

Part 7 Write the missing numbers.

30   29   ___   ___   ___   25   ___   ___   ___   ___   ___   20

Part 8

a. 52	b. 380	c. 109	d. 47
tens digit _____	hundreds digit _____	hundreds digit _____	tens digit _____
ones digit _____	ones digit _____	tens digit _____	ones digit _____

Lesson 9

Part 1

a. $5 + 10 =$ ____

b. $2 + 10 =$ ____

c. $9 + 10 =$ ____

d. $6 + 10 =$ ____

e. $3 + 10 =$ ____

f. $8 + 10 =$ ____

Part 2

a. $1 + 5 =$ ____

h. $5 + 1 =$ ____

o. $9 - 1 =$ ____

b. $10 - 9 =$ ____

i. $10 - 9 =$ ____

p. $1 + 3 =$ ____

c. $5 - 1 =$ ____

j. $3 - 1 =$ ____

q. $1 + 5 =$ ____

d. $1 + 8 =$ ____

k. $8 - 7 =$ ____

r. $6 - 5 =$ ____

e. $6 - 5 =$ ____

l. $6 + 1 =$ ____

s. $1 + 4 =$ ____

f. $9 - 1 =$ ____

m. $1 + 8 =$ ____

t. $3 - 1 =$ ____

g. $1 + 4 =$ ____

n. $10 + 1 =$ ____

Part 3

a.
$$\begin{array}{r} 44 \\ + 32 \\ \hline \end{array}$$

b.
$$\begin{array}{r} 15 \\ + 33 \\ \hline \end{array}$$

c.
$$\begin{array}{r} 56 \\ + 23 \\ \hline \end{array}$$

d.
$$\begin{array}{r} 34 \\ + 13 \\ \hline \end{array}$$

e.
$$\begin{array}{r} 68 \\ + 30 \\ \hline \end{array}$$

f.
$$\begin{array}{r} 13 \\ + 82 \\ \hline \end{array}$$

Lesson 9

Name _____

Part 4

a. 4
  + 2

b. 6
  + 0

c. 6
  + 2

d. 8
  − 8

e. 7
  − 0

f. 9
  − 1

g. 5
  + 1

h. 9
  − 0

i. 10
  + 0

j. 10
  + 2

k. 8
  − 1

l. 7
  − 7

Part 5 Write the missing numbers.

20   19   ___   ___   ___   ___   14   ___   ___   ___   10

Part 6 Write 4 facts for each family.

a. 6 ——— 4→ 10

_____

_____

_____

_____

b. 7 ——— 3→ 10

_____

_____

_____

_____

Part 7

a. 109	b. 38	c. 594	d. 805
hundreds digit _____	tens digit _____	hundreds digit _____	tens digit _____
tens digit _____	ones digit _____	ones digit _____	ones digit _____

Lesson

Name _____

Part 1

a. 4 + 10 = _____   4 + 9 = _____

b. 6 + 10 = _____   6 + 9 = _____

c. 3 + 10 = _____   3 + 9 = _____

d. 8 + 10 = _____   8 + 9 = _____

e. 2 + 10 = _____   2 + 9 = _____

Part 2

a. ■ —1→ 6

_____

b. 5 —■→ 6

_____

c. ■ —1→ 10

_____

d. 9 —■→ 10

_____

e. ■ —1→ 8

_____

f. 7 —■→ 8

_____

Part 3

a. —2→ 7

_____

b. 9 —1→ ___

_____

c. —2→ 11

_____

d. 7 —→ 8

_____

e. 8 —2→ ___

_____

f. 2 —→ 4

_____

Connecting Math Concepts

Lesson

Name _____

a. 10 + 1 = _____   h. 5 – 1 = _____   o. 1 + 8 = _____

b. 5 + 1 = _____   i. 9 – 1 = _____   p. 1 + 4 = _____

c. 3 – 1 = _____   j. 1 + 5 = _____   q. 10 + 1 = _____

d. 6 – 5 = _____   k. 3 – 1 = _____   r. 1 + 5 = _____

e. 10 – 9 = _____   l. 10 – 9 = _____   s. 8 – 7 = _____

f. 1 + 4 = _____   m. 6 + 1 = _____   t. 6 – 5 = _____

g. 1 + 8 = _____   n. 9 – 1 = _____

Part 5

a.   2 3   b.   1 7   c.   4 3
   + 5 2   + 1 2   + 1 6

d.   3 8   e.   1 5   f.   3 4
   + 3 0   + 6 3   + 3 3

Independent Work

Part 6

a. 328	b. 108	c. 45	d. 328
tens digit _____	ones digit _____	tens digit _____	hundreds digit _____
ones digit _____	hundreds digit _____	ones digit _____	tens digit _____

Lesson 10

Name _____

Part 7

a.	b.	c.	d.	e.	f.
10 + 0	8 + 0	8 + 2	6 − 6	5 − 0	7 − 1

g.	h.	i.	j.	k.	l.
7 + 1	7 − 0	4 + 0	4 + 2	6 − 1	9 − 9

Part 8 Write the missing numbers.

20   19   ___   ___   ___   15   ___   ___   ___   ___   ___

Part 9 Write 4 facts for each family.

a.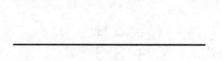
5   3   8

_____

_____

_____

_____

b. 3   4   7

_____

_____

_____

_____

Part 10

a.	b.	c.	d.	e.	f.
8 + 1	4 + 3	7 + 0	7 + 3	9 + 2	2 + 3

g.	h.	i.	j.	k.	l.
5 + 3	5 + 2	8 + 3	6 + 2	9 + 1	4 + 3

Connecting Math Concepts

Lesson

Name _____

a. ■ —1→ 7   b. 6 ■ —→ 7

_____   _____

c. ■ —1→ 9   d. 8 ■ —→ 9

_____   _____

e. ■ —1→ 4   f. 3 ■ —→ 4

_____   _____

g. ■ —1→ 8   h. 7 ■ —→ 8

_____   _____

Part 2

a. _____ 37   b. _____ 60

c. _____ 12   d. _____ 10

Part 3

a. 37 is more than 29.   b. 93 is less than 100.

c. 23 is less than 24.   d. 38 is more than 21.

Lesson 11

Name _____

Part 4

a. $7 + 10 =$ _____ $7 + 9 =$ _____

b. $4 + 10 =$ _____ $4 + 9 =$ _____

c. $9 + 10 =$ _____ $9 + 9 =$ _____

d. $5 + 10 =$ _____ $5 + 9 =$ _____

e. $3 + 10 =$ _____ $3 + 9 =$ _____

Part 5

a. $5 \xrightarrow{\qquad} 6$

_____

b. $8 \xrightarrow{\quad 2 \quad}$ ___

_____

c. $\xrightarrow{\quad 1 \quad} 10$

_____

d. $6 \xrightarrow{\quad 2 \quad}$ ___

_____

Part 6

a. $8 - 1 =$ _____   h. $4 + 1 =$ _____   o. $9 - 8 =$ _____

b. $1 + 10 =$ _____   i. $7 - 1 =$ _____   p. $6 - 5 =$ _____

c. $1 + 6 =$ _____   j. $7 - 6 =$ _____   q. $9 + 1 =$ _____

d. $3 - 2 =$ _____   k. $1 + 9 =$ _____   r. $1 + 6 =$ _____

e. $11 - 10 =$ _____   l. $1 + 7 =$ _____   s. $3 - 2 =$ _____

f. $7 - 1 =$ _____   m. $8 - 1 =$ _____   t. $10 - 1 =$ _____

g. $2 - 1 =$ _____   n. $11 - 10 =$ _____

Connecting Math Concepts

Lesson 11

Independent Work

Part 7

a. 38

ones digit _____

tens digit _____

b. 305

tens digit _____

hundreds digit _____

c. 148

hundreds digit _____

tens digit _____

d. 10

tens digit _____

ones digit _____

Part 8

20 _____ _____ _____ 16 _____ _____ _____ 12 _____ _____

Part 9

a. 1
  + 7

b. 8
  − 8

c. 9
  − 0

d. 1
  + 4

e. 10
  + 0

f. 5
  + 3

g. 7
  − 1

h. 1
  + 3

i. 6
  − 6

j. 4
  − 1

k. 1
  + 6

l. 3
  + 3

Part 10 Write 4 facts for each family.

a. 4 ———→ 3 ——→ 7

_____

_____

_____

_____

b. 5 ———→ 3 ——→ 8

_____

_____

_____

_____

Part 11

a. 4 3
  + 2 1

b. 5 5
  + 3 0

c. 2 6
  + 1 3

d. 1 5
  + 6 2

Lesson

Name _____

Part 1

a. $\underline{6} \longrightarrow 8$

_____

b. $\underline{} \xrightarrow{1} 5$

_____

c. $\underline{7 \qquad 2} \longrightarrow \underline{}$

_____

d. $\underline{} \xrightarrow{2} 10$

_____

e. $\underline{5} \longrightarrow 7$

_____

f. $\underline{9} \longrightarrow 11$

_____

Part 2

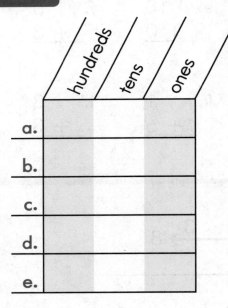

Part 3

a. 8 + 10 = _____   8 + 9 = _____

b. 6 + 10 = _____   6 + 9 = _____

c. 1 + 10 = _____   1 + 9 = _____

d. 7 + 10 = _____   7 + 9 = _____

e. 3 + 10 = _____   3 + 9 = _____

Part 4

a. 44 is more than K.

b. 70 is more than B.

c. K is more than 70.

Lesson

Part 5

a. _____ 16   b. _____ 20

c. _____ 10   d. _____ 96

Part 6

a. ■ ——2——▶ 8   b. 6 ——■——▶ 8

_____   _____

c. ■ ——2——▶ 5   d. 3 ——■——▶ 5

_____   _____

e. ■ ——2——▶ 12   f. 10 ——■——▶ 12

_____   _____

Part 7

a. 11 − 1 = _____   h. 2 + 1 = _____   o. 7 − 6 = _____

b. 4 − 3 = _____   i. 4 − 3 = _____   p. 1 + 9 = _____

c. 1 + 9 = _____   j. 5 − 4 = _____   q. 4 − 1 = _____

d. 7 − 6 = _____   k. 3 + 1 = _____   r. 1 + 2 = _____

e. 5 − 4 = _____   l. 11 − 1 = _____   s. 11 − 10 = _____

f. 4 − 1 = _____   m. 2 − 1 = _____   t. 8 + 1 = _____

g. 8 − 7 = _____   n. 7 + 1 = _____

Lesson 12

Independent Work

Part 8

a. 35 + 10 = _____   d. 21 + 10 = _____

b. 72 + 10 = _____   e. 18 + 10 = _____

c. 50 + 10 = _____

Part 9

a. 312	b. 78	c. 609	d. 81
hundreds digit _____	ones digit _____	tens digit _____	tens digit _____
tens digit _____	tens digit _____	hundreds digit _____	ones digit _____

Part 10

a. $\begin{array}{r} 9 \\ -9 \\ \hline \end{array}$ b. $\begin{array}{r} 1 \\ +8 \\ \hline \end{array}$ c. $\begin{array}{r} 2 \\ -0 \\ \hline \end{array}$ d. $\begin{array}{r} 7 \\ +2 \\ \hline \end{array}$ e. $\begin{array}{r} 1 \\ +3 \\ \hline \end{array}$ f. $\begin{array}{r} 5 \\ +3 \\ \hline \end{array}$

g. $\begin{array}{r} 8 \\ -3 \\ \hline \end{array}$ h. $\begin{array}{r} 6 \\ -6 \\ \hline \end{array}$ i. $\begin{array}{r} 9 \\ +0 \\ \hline \end{array}$ j. $\begin{array}{r} 1 \\ +7 \\ \hline \end{array}$ k. $\begin{array}{r} 10 \\ -10 \\ \hline \end{array}$ l. $\begin{array}{r} 5 \\ -1 \\ \hline \end{array}$

Part 11

a. $\begin{array}{r} 41 \\ +17 \\ \hline \end{array}$ b. $\begin{array}{r} 53 \\ +22 \\ \hline \end{array}$ c. $\begin{array}{r} 86 \\ +10 \\ \hline \end{array}$ d. $\begin{array}{r} 34 \\ +32 \\ \hline \end{array}$

Part 12 Write the missing numbers.

_____ 19 _____ _____ _____ _____ _____ _____ _____ 11 _____

Lesson 13

Name _____

a. $\xrightarrow{\quad 2 \quad}$ 12

_____

d. $\xrightarrow{\quad 2 \quad}$ 10

_____

g. $\xrightarrow{\quad 2 \quad}$ 5

_____

b. 8 $\xrightarrow{\quad\quad}$ 10

_____

e. 7 $\xrightarrow{\quad 1 \quad}$ __

_____

h. 3 $\xrightarrow{\quad\quad}$ 4

_____

c. 8 $\xrightarrow{\quad 1 \quad}$ __

_____

f. 6 $\xrightarrow{\quad\quad}$ 8

_____

Part 2

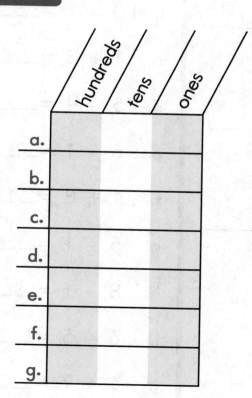

Part 3

a. 86 is less than C.

b. R is more than 51.

c. 45 is less than K.

d. J is more than 66.

Part 4

a. $6 + 9 =$ _____   f. $5 + 9 =$ _____

b. $8 + 9 =$ _____   g. $1 + 9 =$ _____

c. $2 + 9 =$ _____   h. $3 + 9 =$ _____

d. $9 + 9 =$ _____   i. $4 + 9 =$ _____

e. $7 + 9 =$ _____

Part 5

a. ■ —2→ 10   b. 8 —■→ 10

_____   _____

c. ■ —2→ 7   d. 5 —■→ 7

_____   _____

e. ■ —2→ 9   f. 7 —■→ 9

_____   _____

Part 6

a. $8 - 7 =$ ____

b. $10 - 1 =$ ____

c. $1 + 3 =$ ____

d. $5 - 4 =$ ____

e. $6 - 1 =$ ____

f. $1 + 8 =$ ____

g. $5 - 1 =$ ____

h. $8 - 1 =$ ____

i. $6 - 5 =$ ____

j. $1 + 7 =$ ____

Lesson 13

Independent Work

Part 7

a.
```
  5 8
– 1 2
```

b.
```
  4 9
– 3 0
```

c.
```
  6 5
– 2 1
```

d.
```
  7 7
– 2 0
```

Part 8 Write the place-value facts.

a. _____ = 13   b. _____ = 30

c. _____ = 15   d. _____ = 50

Part 9 Write the missing numbers.

26 ___ ___ 24 ___ ___ ___ ___ 20 ___

Part 10

a. 40 + 10 = ____   d. 51 + 10 = ____

b. 35 + 10 = ____   e. 89 + 10 = ____

c. 17 + 10 = ____

Lesson

Name _____

a. $\underline{8} \longrightarrow 10$

_____

b. $\xrightarrow{\quad 1 \quad} 9$

_____

c. $\underline{6 \quad\quad 2} \longrightarrow \underline{\quad}$

_____

d. $\underline{4} \longrightarrow 5$

_____

e. $\xrightarrow{\quad 2 \quad} 10$

_____

f. $\underline{8 \quad\quad 1} \longrightarrow \underline{\quad}$

_____

g. $\underline{10} \longrightarrow 11$

_____

h. $\underline{7} \longrightarrow 9$

_____

Part 2

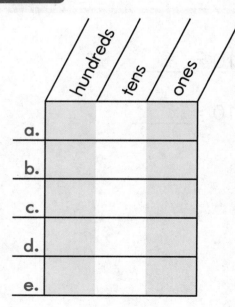

Part 3

a. 34 is less than M.

b. K is less than 16.

c. R is more than 70.

d. 46 is less than T.

e. V is less than 9.

Connecting Math Concepts

Lesson

Name _____

Part 4

a. $7 + 9 =$ _____

b. $3 + 9 =$ _____

c. $8 + 9 =$ _____

d. $6 + 9 =$ _____

e. $9 + 9 =$ _____

f. $1 + 9 =$ _____

g. $4 + 9 =$ _____

h. $2 + 9 =$ _____

i. $5 + 9 =$ _____

Part 5

a. A penny is worth 1 cent.

b. A nickel is worth 5 cents.

c. A dime is worth 10 cents.

d. A quarter is worth 25 cents.

Part 6

a. ▢ + ▢ + ▢ = 563

b. ▢ + ▢ + ▢ = 492

c. ▢ + ▢ + ▢ = 730

d. ▢ + ▢ + ▢ = 178

Part 7

a. $4 - 1 =$ _____

b. $7 - 6 =$ _____

c. $1 + 5 =$ _____

d. $10 - 9 =$ _____

e. $11 - 1 =$ _____

f. $9 - 8 =$ _____

g. $1 + 8 =$ _____

h. $3 - 2 =$ _____

i. $1 + 10 =$ _____

j. $9 - 1 =$ _____

Lesson 14

Independent Work

Part 8 Write the place-value facts.

a. _____ = 10   b. _____ = 14

c. _____ = 40   d. _____ = 27

Part 9

a. 6 3   b. 7 8   c. 4 5   d. 5 6
  − 2 2   − 1 0   − 1 2   − 2 1

Part 10 Write the missing numbers.

__25__ ___ __23__ ___ __21__ ___ ___ ___ __17__

Part 11 Write the subtraction facts.

a. ■ —2→ 8   b. 6 —■→ 8

_____   _____

c. ■ —2→ 5   d. 3 —■→ 5

_____   _____

e. ■ —2→ 9   f. 7 —■→ 9

_____   _____

Connecting Math Concepts

Lesson

Name _____

a. K is more than 21.

■ ——————→

b. 16 is less than P.

■ ——————→

c. J is less than 15.

■ ——————→

d. 13 is more than B.

■ ——————→

Part 2

a. penny _____ cents

b. nickel _____ cents

c. dime _____ cents

d. quarter _____ cents

Part 3

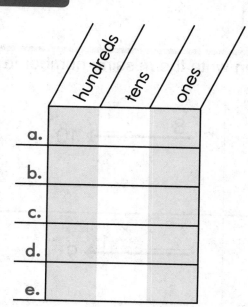

	hundreds	tens	ones
a.			
b.			
c.			
d.			
e.			

Part 4

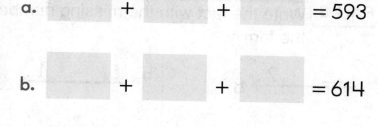

a. ____ + ____ + ____ = 593

b. ____ + ____ + ____ = 614

c. ____ + ____ + ____ = 476

d. ____ + ____ + ____ = 820

Lesson 15

Name _____

Part 5

a. $3 - 1 = $____ | i. $11 - 1 = $____ | q. $8 - 7 = $____ | x. $5 - 4 = $____

b. $1 + 2 = $____ | j. $1 + 3 = $____ | r. $7 - 1 = $____ | y. $9 + 1 = $____

c. $3 - 2 = $____ | k. $10 - 9 = $____ | s. $1 + 8 = $____ | z. $5 - 1 = $____

d. $6 - 1 = $____ | l. $4 - 1 = $____ | t. $4 - 3 = $____ | A. $8 + 1 = $____

e. $1 + 5 = $____ | m. $1 + 9 = $____ | u. $1 + 7 = $____ | B. $7 - 6 = $____

f. $9 - 8 = $____ | n. $6 - 5 = $____ | v. $8 - 1 = $____ | C. $9 - 1 = $____

g. $10 - 1 = $____ | o. $2 - 1 = $____ | w. $11 - 10 = $____ | D. $1 + 6 = $____

h. $1 + 10 = $____ | p. $1 + 4 = $____

Independent Work

Part 6 Write the fact with the missing number. Then write the missing number in the family.

a. ___ $\xrightarrow{2}$ 8

b. $5 \xrightarrow{1}$ ___

c. $8 \xrightarrow{}$ 10

_____

_____

_____

d. $6 \xrightarrow{}$ 7

e. $4 \xrightarrow{2}$ ___

f. ___ $\xrightarrow{1}$ 9

_____

_____

_____

Connecting Math Concepts

Lesson 15

Name _____

Copyright © The McGraw-Hill Companies, Inc.

Part 7 Write the place-value facts.

a. _____ = 27   d. _____ = 13

b. _____ = 72   e. _____ = 31

c. _____ = 20   f. _____ = 70

Part 8 Write the missing numbers.

__25__ ___ ___ ___ ___ ___ __19__ ___ ___ ___

Part 9

a. 24   b. 52   c. 16   d. 46   e. 91
   +33   −10   +13   +32   −20

Part 10 Write the subtraction facts.

a. ■ ——→ 2 ——→ 3   b. 1 ——→ ■ ——→ 3

_____   _____

c. ■ ——→ 1 ——→ 8   d. 7 ——→ ■ ——→ 8

_____   _____

e. ■ ——→ 2 ——→ 7   f. 5 ——→ ■ ——→ 7

_____   _____

Connecting Math Concepts

Lesson 16

Name _____

Part 1

a. ☐ + ☐ + ☐ = 345

b. ☐ + ☐ + ☐ = 219

c. ☐ + ☐ + ☐ = 530

d. ☐ + ☐ + ☐ = 687

Part 2

a.		
b.		
c.		
d.		
e.		

Part 3

a. 23 is less than Z.

■ ⟶

b. L is less than 7.

■ ⟶

c. P is more than 56.

■ ⟶

d. 6 is more than K.

■ ⟶

Part 4

a. 1 + 6 + 2 = _____

b. 2 + 2 + 3 = _____

c. 7 + 1 + 2 = _____

Part 5

a. 6 − 5 = _____
b. 10 + 1 = _____
c. 1 + 11 = _____
d. 4 − 3 = _____
e. 9 − 1 = _____
f. 1 + 10 = _____
g. 9 − 8 = _____
h. 5 − 1 = _____

i. 2 − 1 = _____
j. 1 + 5 = _____
k. 1 + 7 = _____
l. 8 − 1 = _____
m. 11 − 10 = _____
n. 4 − 1 = _____
o. 1 + 6 = _____
p. 7 − 6 = _____

q. 8 − 7 = _____
r. 3 − 1 = _____
s. 1 + 8 = _____
t. 1 + 6 = _____
u. 7 − 1 = _____
v. 1 + 9 = _____
w. 11 − 1 = _____

x. 3 − 2 = _____
y. 6 − 1 = _____
z. 8 + 1 = _____
A. 5 − 4 = _____
B. 10 − 9 = _____
C. 10 − 1 = _____
D. 1 + 4 = _____

Connecting Math Concepts

Lesson 16

Name _____

Independent Work

Part 6 Write the cents for each row.

a. [coins] ▢ cents

b. [coins] ▢ cents

c. [coins] ▢ cents

Part 7 Write the place-value facts.

a. _____ = 17   d. _____ = 13

b. _____ = 80   e. _____ = 77

c. _____ = 30

Part 8 Write the missing numbers.

43 ___ ___ 40 ___ ___ ___ 36

Part 9

a. $\begin{array}{r} 7 \\ + 9 \\ \hline \end{array}$
b. $\begin{array}{r} 2 \\ + 9 \\ \hline \end{array}$
c. $\begin{array}{r} 5 \\ + 9 \\ \hline \end{array}$
d. $\begin{array}{r} 8 \\ + 9 \\ \hline \end{array}$
e. $\begin{array}{r} 3 \\ + 9 \\ \hline \end{array}$

Lesson 16

Name _____

Part 10 Write the fact. Then write the missing number in the family.

a. 7 —— 3 → ___

_____

b. —— 1 → 7

_____

c. 7 —— → 9

_____

d. 6 —— → 8

_____

e. 5 —— 3 → ___

_____

f. —— 2 → 10

_____

Part 11

a. 1 5
   +2 3

b. 3 6
   –2 0

c. 4 9
   –1 2

d. 5 7
   +3 1

e. 9 9
   –2 2

Connecting Math Concepts

Lesson

Name _____

Part 1

a. ⬜ + ⬜ + ⬜ = 506

b. ⬜ + ⬜ + ⬜ = 380

c. ⬜ + ⬜ + ⬜ = 809

d. ⬜ + ⬜ + ⬜ = 300

e. ⬜ + ⬜ + ⬜ = 805

Part 2

a. 40 ____ 10 → G

_____

b. 7 ____ W → 9

_____

Part 3

a.  ⬜ cents

b. ⬜ cents

c. ⬜ cents

d.  ⬜ cents

Lesson 17

Name _____

Part 4

a. $1 + 4 + 1 =$ _____

b. $8 + 1 + 2 =$ _____

c. $2 + 2 + 3 =$ _____

d. $8 + 2 + 6 =$ _____

Part 5

a. $2 + 5 =$ _____ k. $10 - 8 =$ _____

b. $7 - 2 =$ _____ l. $7 - 2 =$ _____

c. $7 - 5 =$ _____ m. $7 - 5 =$ _____

d. $2 + 8 =$ _____ n. $2 + 5 =$ _____

e. $10 - 2 =$ _____ o. $10 - 2 =$ _____

f. $10 - 8 =$ _____ p. $6 - 4 =$ _____

g. $2 + 4 =$ _____ q. $2 + 4 =$ _____

h. $6 - 2 =$ _____ r. $6 + 2 =$ _____

i. $6 - 4 =$ _____ s. $6 - 2 =$ _____

j. $4 + 2 =$ _____ t. $2 + 8 =$ _____

Independent Work

Part 6 Write the missing numbers.

a. __5__ ___ ___ __20__ ___ ___ ___ ___ ___ __50__

b. __2__ ___ ___ ___ __10__ ___ ___ ___ __18__ ___

Connecting Math Concepts

Lesson 17

Name _____

Part 7 Write the missing numbers.

<u>72</u>   <u>71</u>   ___   ___   <u>68</u>   ___   ___   ___   ___

Part 8 Write the place-value facts.

a. _____ = 81   d. _____ = 33

b. _____ = 49   e. _____ = 13

c. _____ = 18

Part 9

a. $\begin{array}{r} 3\,5 \\ +\,1\,2 \\ \hline \end{array}$ b. $\begin{array}{r} 4\,1 \\ +\,3\,7 \\ \hline \end{array}$ c. $\begin{array}{r} 5\,8 \\ -\,2\,1 \\ \hline \end{array}$ d. $\begin{array}{r} 9\,5 \\ -\,1\,1 \\ \hline \end{array}$ e. $\begin{array}{r} 1\,6 \\ +\,3\,0 \\ \hline \end{array}$

Part 10

a. $19 + 10 =$ _____   b. $50 + 10 =$ _____   c. $16 + 10 =$ _____

d. $34 + 10 =$ _____   e. $82 + 10 =$ _____   f. $61 + 10 =$ _____

Part 11

a. $6 + 1 =$ ____   d. $8 - 1 =$ ____   g. $11 + 1 =$ ____   j. $9 - 8 =$ ____

b. $5 - 4 =$ ____   e. $1 + 5 =$ ____   h. $10 - 1 =$ ____   k. $7 + 1 =$ ____

c. $2 - 1 =$ ____   f. $7 - 6 =$ ____   i. $1 + 4 =$ ____   l. $6 - 1 =$ ____

Connecting Math Concepts   Lesson 17   **37**

Lesson

Part 1

a. ☐ + ☐ + ☐ = 100

b. ☐ + ☐ + ☐ = 472

c. ☐ + ☐ + ☐ = 703

d. ☐ + ☐ = 60

Part 2

a. V $\xrightarrow{\quad 2 \quad}$ 9

_____

b. 10 $\xrightarrow{\quad 5 \quad}$ D

_____

c. 30 $\xrightarrow{\quad 7 \quad}$ F

_____

d. 9 $\xrightarrow{\quad N \quad}$ 10

_____

Part 3

a. 2 + 7 = _____

b. 9 – 2 = _____

c. 9 – 7 = _____

d. 2 + 3 = _____

e. 5 – 2 = _____

f. 5 – 3 = _____

g. 2 + 10 = _____

h. 12 – 2 = _____

i. 12 – 10 = _____

j. 10 + 2 = _____

k. 9 – 7 = _____

l. 2 + 7 = _____

m. 5 – 2 = _____

n. 2 + 3 = _____

o. 9 – 2 = _____

p. 2 + 10 = _____

q. 12 – 10 = _____

r. 5 – 3 = _____

s. 12 – 2 = _____

t. 3 + 2 = _____

Connecting Math Concepts

Lesson

Name _____

Part 4

a. $5 + 2 + 1 =$ _____

b. $9 + 1 + 4 =$ _____

c. $2 + 3 + 1 =$ _____

d. $1 + 4 + 3 =$ _____

e. $2 + 2 + 2 =$ _____

f. $7 + 2 + 1 =$ _____

Part 5 Write the cents for each row.

a.   cents

b.   cents

c.   cents

Independent Work

Part 6 Write the missing numbers.

a. 100 ___ ___ 70 ___ ___ ___ ___ ___ 10

b. 83 82 ___ ___ ___ 78 ___ ___ ___ ___

Part 7

a. $\begin{array}{r} 48 \\ +21 \\ \hline \end{array}$

b. $\begin{array}{r} 13 \\ +52 \\ \hline \end{array}$

c. $\begin{array}{r} 45 \\ -12 \\ \hline \end{array}$

d. $\begin{array}{r} 88 \\ -20 \\ \hline \end{array}$

e. $\begin{array}{r} 66 \\ +32 \\ \hline \end{array}$

Lesson 18

Name _____

Part 8 Write the missing numbers.

a. __2__ ___ ___ ___ ___ __12__ ___ ___ ___ ___

b. __5__ ___ ___ ___ ___ ___ __40__ ___ ___

Part 9

a.	b.	c.	d.	e.
4 + 9	4 + 3	7 + 3	6 + 9	7 + 9

Part 10 Write the fact. Then write the missing number in the family.

a. ___ ——2→ 8

_____

b. 4 ——3→ ___

_____

c. ___ ——1→ 7

_____

d. 8 ——→ 9

_____

e. 9 ——→ 11

_____

f. 5 ——3→ ___

_____

Part 11

a. $8 - 1 =$ ____   d. $9 + 1 =$ ____   g. $7 - 6 =$ ____   j. $1 + 10 =$ ____

b. $1 + 5 =$ ____   e. $8 - 7 =$ ____   h. $11 - 1 =$ ____   k. $9 - 8 =$ ____

c. $3 - 1 =$ ____   f. $7 + 1 =$ ____   i. $6 - 5 =$ ____   l. $1 + 8 =$ ____

Lesson

Name _____

Part 1

a. _____

b. _____

c. _____

d. _____

e. _____

Part 2

a. 8 ——— 3→ P

_____

b. 6 ——— W→ 8

_____

c. V ——— 9→ 12

_____

d. 9 ——— 9→ Y

_____

Part 3

a. 2 + 6 = _____   k. 11 − 9 = _____

b. 8 − 2 = _____   l. 8 − 2 = _____

c. 8 − 6 = _____   m. 2 + 2 = _____

d. 2 + 4 = _____   n. 2 + 4 = _____

e. 6 − 2 = _____   o. 11 − 2 = _____

f. 6 − 4 = _____   p. 8 − 6 = _____

g. 2 + 9 = _____   q. 2 + 9 = _____

h. 11 − 2 = _____   r. 2 + 6 = _____

i. 11 − 9 = _____   s. 6 − 4 = _____

j. 9 + 2 = _____   t. 6 − 2 = _____

Part 4

a.   4   b.   6   c.   6   d.   2   e.   7   f.   1
     2   2   2   2   3   8
   + 1   + 10   + 1   + 3   + 6   + 2

Lesson 19

Name _____

Independent Work

Part 5 Write the cents for each row.

a.   [ ] cents

b.    [ ] cents

c.      [ ] cents

---

Part 6 Write the missing numbers.

a. __2__ ___ ___ ___ ___ ___ __14__ ___ ___ ___

b. __5__ ___ ___ __20__ ___ ___ ___ ___ ___

c. __100__ ___ ___ __60__ ___ ___ __30__ ___ ___

Part 7 Write the place-value facts.

a. _____ = 19

b. _____ = 53

Part 8

a. 2 6
  – 1 0
  _____

b. 1 4
  + 2 3
  _____

Part 9 Write the missing numbers.

a. 58 + 10 = _____   **b.** 40 + 10 = _____   **c.** 55 + 10 = _____

d. 13 + 10 = _____   **e.** 81 + 10 = _____   **f.** 32 + 10 = _____

42 Lesson 19   **Connecting Math Concepts**

Lesson

Name _____

Part 1

a. _____

b. _____

c. _____

d. _____

e. _____

Part 2

a. $\underrightarrow{7 \qquad 3}$ H

b. $\underrightarrow{T \qquad 2}$ 12

c. $\underrightarrow{8 \qquad R}$ 10

d. $\underrightarrow{40 \qquad 2}$ F

e. $\underrightarrow{4 \qquad Q}$ 5

Part 3

a. 8 − 6 =_____

b. 10 − 2 =_____

c. 2 + 3 =_____

d. 11 − 2 =_____

e. 6 − 4 =_____

f. 2 + 5 =_____

g. 9 − 7 =_____

h. 2 + 10 =_____

i. 5 − 2 =_____

j. 10 − 8 =_____

Part 4

a.
```
   5
   2
 + 1
```

b.
```
   9
   1
 + 7
```

c.
```
   3
   3
 + 2
```

d.
```
   7
   2
 + 1
```

e.
```
   8
   2
 + 3
```

f.
```
   1
   4
 + 2
```

Lesson 20

Name _____

Independent Work

Part 5 Write the cents for each row.

a.   ☐ cents

b.   ☐ cents

c.   ☐ cents

Part 6 Write the missing numbers. | **Part 7**

a. ____9____

____18____

_____

_____

____45____

_____

____63____

_____

_____

____90____

Part 7

a. 4 3
  − 1 0

b. 5 2
  + 3 2

c. 8 5
  − 1 2

d. 9 7
  − 2 1

e. 5 4
  + 2 1

Part 8 Write the missing numbers.

a. __100__ ____ ____ __70__ ____ ____ ____ ____ __20__ ____

b. __2__ ____ ____ ____ ____ ____ ____ __18__ ____

Connecting Math Concepts

Lesson 20

Name _____

Part 9 Write the fact. Then write the missing number in the family.

a. 7 ———3→ ___

_____

b. ══ ——1→ 6

_____

c. 4 ———3→ ___

_____

d. 7 ———→ 8

_____

e. ══ ——2→ 5

_____

f. 8 ———→ 10

_____

Part 10

a.   6
   + 9

b.   5
   + 3

c.   7
   + 9

d.   3
   + 9

e.   8
   + 3

Lesson 21

Name _____

Part 1

a. _____

b. _____

c. _____

d. _____

e. _____

Part 2

a. $2 + 4 =$ _____

b. $7 - 5 =$ _____

c. $11 - 9 =$ _____

d. $8 - 2 =$ _____

e. $12 - 2 =$ _____

f. $2 + 6 =$ _____

g. $5 - 3 =$ _____

h. $2 + 7 =$ _____

i. $12 - 10 =$ _____

j. $9 - 2 =$ _____

Independent Work

Part 3

a.
$$\begin{array}{r} 8 \\ 1 \\ + 3 \\ \hline \end{array}$$

b.
$$\begin{array}{r} 7 \\ 3 \\ + 5 \\ \hline \end{array}$$

c.
$$\begin{array}{r} 4 \\ 9 \\ + 1 \\ \hline \end{array}$$

d.
$$\begin{array}{r} 3 \\ 3 \\ + 3 \\ \hline \end{array}$$

e.
$$\begin{array}{r} 5 \\ 2 \\ + 3 \\ \hline \end{array}$$

f.
$$\begin{array}{r} 1 \\ 9 \\ + 6 \\ \hline \end{array}$$

Connecting Math Concepts

Lesson 21

Name _____

Part 4 Write the missing numbers.

a. ___9___

   _____

   ___27___

   _____

   ___45___

   _____

   ___63___

   _____

   ___81___

   _____

Part 5

a.   9 5
   – 1 2

b.   5 7
   – 2 0

c.   4 8
   + 2 1

Part 6

a. 1 + 0 = _____   d. 5 + 1 = _____   g. 5 – 1 = _____   j. 3 – 1 = _____

b. 6 – 1 = _____   e. 8 – 1 = _____   h. 1 + 9 = _____   k. 1 + 3 = _____

c. 4 – 3 = _____   f. 1 + 7 = _____   i. 10 – 9 = _____   l. 1 – 1 = _____

Part 7 Write the place-value facts.

a. _____ = 17   b. _____ = 33

c. _____ = 50   d. _____ = 91

Lesson 22

Name _____

Part 1

a. 15 is 90 less than K.

b. P is 22 more than 39.

c. K is 9 less than 14.

d. 28 is 12 more than B.

Part 2

a. 4 – 2 = _____

b. 8 + 2 = _____

c. 11 – 9 = _____

d. 5 + 2 = _____

e. 5 – 3 = _____

f. 12 – 2 = _____

g. 8 – 6 = _____

h. 7 + 2 = _____

i. 6 – 4 = _____

j. 2 + 2 = _____

k. 10 – 2 = _____

l. 12 – 10 = _____

m. 9 – 7 = _____

n. 6 – 2 = _____

o. 4 + 2 = _____

p. 7 – 5 = _____

q. 5 – 2 = _____

r. 9 – 2 = _____

s. 7 – 2 = _____

t. 3 + 2 = _____

u. 11 – 2 = _____

v. 9 + 2 = _____

w. 8 – 2 = _____

x. 10 + 2 = _____

y. 3 + 2 = _____

z. 10 – 8 = _____

A. 7 – 5 = _____

B. 6 + 2 = _____

C. 11 – 2 = _____

D. 11 – 9 = _____

Part 3

a. $\dfrac{5 \quad\quad C}{} \longrightarrow$ M

M = 7

b. $\dfrac{20 \quad\quad N}{} \longrightarrow$ K

N = 6

Connecting Math Concepts

Lesson 22

Name _____

Part 4 Complete the place-value facts.

a. _____ | 205 |

b. _____ | 514 |

c. _____ | 40 |

d. _____ | 780 |

Part 5

a. 136 + 10 = _____

b. 132 + 10 = _____

c. 130 + 10 = _____

d. 137 + 10 = _____

Part 6 Write the fact. Then write the missing number in the family.

a. 10 ———P——→ 11

_____

b. G ———2——→ 9

_____

c. 10 ———8——→ V

_____

d. 20 ———7——→ Y

_____

Part 7 Write the missing numbers.

a. 9 ___ ___ 36 ___ 54 ___ ___ 81 ___

b. 100 ___ ___ ___ 50 ___ ___ ___ ___

Part 8

a. 8
    2
  + 5
  ____

b. 1
    4
  + 9
  ____

c. 7
    1
  + 3
  ____

d. 2
    2
  + 2
  ____

Lesson 23

Name _____

Part 1

a. D is 72 less than 83.

b. 49 is 20 more than K.

c. 16 is 11 less than V.

→   →   →

Part 2

a.   19   12   16

b.   17   9   2

c.   23   30   20

___ ___ ___   ___ ___ ___   ___ ___ ___

Part 3

a.   ☐
   3 4
  + 3 9

b.   ☐
   5 6
  + 2 9

c.   ☐
   4 9
  + 1 3

Part 4

a.   cents

b.   cents

c.   cents

Lesson

Part 5

a. $\dfrac{8 \qquad B}{} \to M$

$\boxed{B = 3}$

b. $\dfrac{P \qquad 8}{} \to G$

$\boxed{G = 10}$

c. $\dfrac{2 \qquad L}{} \to J$

$\boxed{J = 9}$

d. $\dfrac{T \qquad 8}{} \to K$

$\boxed{T = 40}$

Part 6

a. $8 - 2 =$ ____	i. $12 - 10 =$ ____	q. $6 + 2 =$ ____	x. $2 + 8 =$ ____
b. $7 + 2 =$ ____	j. $8 - 6 =$ ____	r. $11 - 9 =$ ____	y. $12 - 2 =$ ____
c. $7 - 5 =$ ____	k. $10 + 2 =$ ____	s. $8 + 2 =$ ____	z. $10 - 2 =$ ____
d. $9 + 2 =$ ____	l. $5 - 3 =$ ____	t. $7 - 2 =$ ____	A. $2 + 2 =$ ____
e. $6 - 4 =$ ____	m. $9 - 7 =$ ____	u. $2 + 9 =$ ____	B. $5 - 2 =$ ____
f. $9 - 2 =$ ____	n. $4 + 2 =$ ____	v. $6 - 2 =$ ____	C. $9 - 2 =$ ____
g. $10 - 8 =$ ____	o. $3 + 2 =$ ____	w. $4 - 2 =$ ____	D. $2 + 7 =$ ____
h. $5 + 2 =$ ____	p. $11 - 2 =$ ____		

Part 7   Complete the place-value facts.

a. _____ $= 324$

b. $600 + 80 + 6 =$ _____

c. _____ $= 609$

d. $100 + 0 + 6 =$ _____

Lesson 23

Independent Work

Part 8

a. $8 + 2 + 5 =$ _____

b. $2 + 7 + 2 =$ _____

c. $16 + 1 + 10 =$ _____

d. $4 + 10 + 10 =$ _____

Part 9

a. $173 + 10 =$ _____

b. $175 + 10 =$ _____

c. $179 + 10 =$ _____

d. $170 + 10 =$ _____

Part 10   Write the fact. Then write the missing number in the family.

a. $\xrightarrow[]{6 \qquad M} 8$

_____

b. $\xrightarrow[]{10 \qquad 7} R$

_____

c. $\xrightarrow[]{V \qquad 10} 11$

_____

Part 11   Write the missing numbers.

a. _____ 9

_____

_____

_____ 36

_____

_____ 63

_____

_____

Part 12

a.
$$\begin{array}{r} 4 \\ + 9 \\ \hline \end{array}$$

b.
$$\begin{array}{r} 7 \\ + 9 \\ \hline \end{array}$$

c.
$$\begin{array}{r} 2 \\ + 9 \\ \hline \end{array}$$

d.
$$\begin{array}{r} 8 \\ + 9 \\ \hline \end{array}$$

Lesson 24

Name _____

Part 1

a. 25 21 20	**b.** 60 90 70	**c.** 10 8 12
___ ___ ___	___ ___ ___	___ ___ ___

Part 2

a. ☐
```
  3 4
+ 3 9
```

b. ☐
```
  6 8
+ 2 3
```

c. ☐
```
  5 3
+ 3 9
```

Part 3

a.    _____ cents

b.   _____ cents

c.    _____ cents

Part 4

a. R is 11 less than 45.

 _____

b. 23 is 41 less than F.

 _____

c. D is 13 more than 56.

 _____

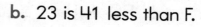

Lesson 24

Name _____

Part 5

a. $11 - 2 =$ ____	i. $11 - 9 =$ ____	q. $2 + 7 =$ ____	x. $3 + 2 =$ ____
b. $6 + 2 =$ ____	j. $10 - 8 =$ ____	r. $10 - 2 =$ ____	y. $9 - 2 =$ ____
c. $5 - 3 =$ ____	k. $7 + 2 =$ ____	s. $2 + 9 =$ ____	z. $8 + 2 =$ ____
d. $10 + 2 =$ ____	l. $6 - 4 =$ ____	t. $9 - 2 =$ ____	A. $12 - 10 =$ ____
e. $8 - 6 =$ ____	m. $7 - 5 =$ ____	u. $2 + 8 =$ ____	B. $5 + 2 =$ ____
f. $7 - 2 =$ ____	n. $9 + 2 =$ ____	v. $5 - 2 =$ ____	C. $4 - 2 =$ ____
g. $9 - 7 =$ ____	o. $2 + 2 =$ ____	w. $8 - 2 =$ ____	D. $9 - 7 =$ ____
h. $4 + 2 =$ ____	p. $12 - 2 =$ ____		

Part 6

a. _____ $= 63$

b. $200 + 0 + 0 =$ _____

c. _____ $= 403$

d. $80 + 9 =$ _____

e. _____ $= 781$

f. _____ $= 571$

g. $100 + 10 + 6 =$ _____

Connecting Math Concepts

Lesson 24

Name _____

Part 7 Complete the place-value facts.

a. _____ = 146

b. $200 + 40 + 8 =$ _____

c. _____ = 609

d. $900 + 10 + 2 =$ _____

Part 8

a.
```
    5
    3
+   2
```

b.
```
    4
    1
+   1
```

c.
```
    1
    9
+   7
```

Part 9 Write the fact. Then write the missing number in the family.

a. $\overset{10 \qquad 3}{\longrightarrow} N$

_____

b. $\overset{2 \qquad Y}{\longrightarrow} 12$

_____

c. $\overset{X \qquad 3}{\longrightarrow} 9$

_____

Part 10

a. _____ 9

_____

_____ 27

_____

_____

_____ 54

_____

_____

_____ 81

_____

Part 11

a. $152 + 10 =$ _____

b. $155 + 10 =$ _____

c. $150 + 10 =$ _____

d. $159 + 10 =$ _____

Lesson

Name _____

Part 1

a. 20 − 10 = _____

b. 17 − 7 = _____

c. 7 + 10 = _____

d. 17 − 10 = _____

e. 18 − 10 = _____

f. 13 − 3 = _____

g. 10 + 10 = _____

h. 18 − 8 = _____

i. 15 − 10 = _____

j. 2 + 10 = _____

Part 2

a. 10 is 17 less than V.

b. K is 13 more than 12.

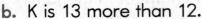

c. P is 21 less than 36.

d. 65 is 55 more than Q.

Part 3

a. 28   18   24

___ ___ ___

b. 35   39   36

___ ___ ___

c. 100   150   50

___ ___ ___

Part 4

a. ☐
  4 9
+ 2 3

b. ☐
  6 7
+ 1 9

c. ☐
  2 8
+ 5 3

Connecting Math Concepts

Lesson 25

Part 5

a. V   40 →F   _____

F = 60

b. 26   P →Y   _____

P = 43

Part 6

a.

_____ dollars   _____ cents

b.

_____ dollars   _____ cents

c.

_____ dollars   _____ cents

Independent Work

Part 7 Complete the place-value facts.

a. 800 + 10 + 0 = _____

b. _____ = 501

c. 200 + 30 + 9 = _____

d. _____ = 636

e. 100 + 0 + 0 = _____

f. _____ = 500

Lesson 25

Part 8

a. 186 + 10 = _____   d. 10 + 56 = _____

b. 126 + 10 = _____   e. 146 + 10 = _____

c. 89 + 10 = _____   f. 143 + 10 = _____

Part 9 Write the missing numbers.

a. _____   b. ___2___

  18   _____

_____   _____

  45   _____

_____   _____

_____   _____

  72   _____

_____   18

_____   _____

Part 10

a.   3 1   b.   7 8   c.   2 1   d.   1 7   e.   8 9

  + 2 6   − 2 0   5 6   4 1   − 1 2

                        + 1 0   + 2 1

Connecting Math Concepts

Lesson 26

Name _____

Part 1

a. K is 17 less than 88.

⟶ _____

b. 14 is 3 less than F.

⟶ _____

c. M is 22 more than 17.

⟶ _____

d. 34 is 10 more than N.

⟶ _____

e. 26 is 30 less than W.

⟶ _____

Part 2

☐
a. 4 7
    + 1 3

☐
b. 2 5
    + 6 9

☐
c. 1 9
    + 7 9

Part 3

a.  _____ dollars _____ cents

b.  _____ dollars _____ cents

c. _____ dollars _____ cents

Lesson 26

Name _____

Part 4

a. $\underrightarrow{\text{K} \qquad 31}$ V   _____

   $\boxed{\text{K} = 45}$

b. $\underrightarrow{\text{R} \qquad \text{T}}$ 78   _____

   $\boxed{\text{R} = 12}$

Part 5

a. $9 + 10 =$ _____

b. $16 - 6 =$ _____

c. $14 - 10 =$ _____

d. $3 + 10 =$ _____

e. $6 + 10 =$ _____

f. $13 - 10 =$ _____

g. $20 - 10 =$ _____

h. $15 - 5 =$ _____

i. $8 + 10 =$ _____

j. $14 - 4 =$ _____

k. $12 - 10 =$ _____

l. $19 - 9 =$ _____

m. $11 - 10 =$ _____

n. $4 + 10 =$ _____

o. $12 - 2 =$ _____

p. $14 - 10 =$ _____

q. $6 + 10 =$ _____

r. $9 + 10 =$ _____

s. $13 - 10 =$ _____

t. $16 - 6 =$ _____

Independent Work

Part 6 Write the 3 numbers from smallest to biggest.

a. 35   38   33

b. 91   92   89

c. 29   25   19

____ ____ ____   ____ ____ ____   ____ ____ ____

Connecting Math Concepts

Lesson 26

Name _____

Part 7

a. 324 + 10 = _____ b. 524 + 10 = _____ c. 146 + 10 = _____

d. 142 + 10 = _____ e. 10 + 83 = _____ f. 10 + 72 = _____

Part 8 Write the fact on the line.

a. $5 \xrightarrow{M} 6$

_____

b. $R \xrightarrow{2} 12$

_____

c. $6 \xrightarrow{3} P$

_____

d. $4 \xrightarrow{N} 6$

_____

Part 9 Write the missing numbers.

a. ___9___

___ ___

___ ___

___ ___

___63___

___ ___

___ ___

___ ___

Part 10 Write the number of cents.

a. ▨ cents

b. ▨ cents

c. ▨ cents

Part 11

a. 8 − 6 = ____ b. 12 − 1 = ____ c. 7 + 2 = ____ d. 2 + 9 = ____

e. 5 + 3 = ____ f. 7 − 2 = ____ g. 2 + 2 = ____ h. 8 + 1 = ____

Lesson 27

Name _____

Part 1

a. N is 15 less than 25.

_____→ _____

b. 26 is 42 less than Y.

_____→ _____

c. T is 22 less than 87.

_____→ _____

d. V is 50 more than 34.

_____→ _____

Part 2

a. ☐ + ☐ = 56

new ☐ + ☐ = 56

b. ☐ + ☐ = 82

new ☐ + ☐ = 82

Part 3

a. $5 + 10 =$ _____

b. $19 - 10 =$ _____

c. $7 + 10 =$ _____

d. $16 - 10 =$ _____

e. $11 - 1 =$ _____

f. $18 - 10 =$ _____

g. $20 - 10 =$ _____

h. $17 - 7 =$ _____

i. $10 + 4 =$ _____

j. $12 - 2 =$ _____

k. $18 - 8 =$ _____

l. $2 + 10 =$ _____

m. $10 + 10 =$ _____

n. $13 - 3 =$ _____

o. $17 - 10 =$ _____

p. $20 - 10 =$ _____

q. $19 - 10 =$ _____

r. $5 + 10 =$ _____

s. $11 - 1 =$ _____

t. $18 - 8 =$ _____

Connecting Math Concepts

Lesson 27

Part 4

a. $\dfrac{62 \qquad K}{}\rightarrow$ P

P = 89

_____

b. $\dfrac{R \qquad 35}{}\rightarrow$ M

R = 32

_____

c. $\dfrac{G \qquad B}{}\rightarrow$ 98

B = 70

_____

Part 5

a. ☐

```
  1 9
+ 6 5
```

b. ☐

```
  2 6
+ 5 3
```

c. ☐

```
  8 4
+ 1 2
```

d. ☐

```
  4 7
+ 3 3
```

Part 6

a.

_____ dollars _____ cents

b.

_____ dollars _____ cents

c.

_____ dollars _____ cents

Lesson 27

Name _____

Part 7 Write the 3 numbers from smallest to biggest.

a. 26   19   30   b. 41   44   40   c. 49   52   51

____ ____ ____   ____ ____ ____   ____ ____ ____

Part 8

a. $1 + 9 =$ ____   d. $9 - 7 =$ ____   g. $6 - 2 =$ ____   i. $10 - 8 =$ ____

b. $2 + 8 =$ ____   e. $6 - 4 =$ ____   h. $8 - 2 =$ ____   j. $5 - 4 =$ ____

c. $13 - 1 =$ ____   f. $2 + 9 =$ ____

Part 9

a.
$$\begin{array}{r} 57 \\ -46 \\ \hline \end{array}$$

b.
$$\begin{array}{r} 53 \\ -33 \\ \hline \end{array}$$

c.
$$\begin{array}{r} 64 \\ 21 \\ +\ \ 3 \\ \hline \end{array}$$

d.
$$\begin{array}{r} 84 \\ -21 \\ \hline \end{array}$$

e.
$$\begin{array}{r} 14 \\ 50 \\ +23 \\ \hline \end{array}$$

Part 10 Write the number of cents.

a.    ▢ cents

b.   ▢ cents

c.   ▢ cents

Part 11 Complete each place-value fact.

a. _____ $= 34$ d. $200 + 0 + 9 =$ ____

b. _____ $= 521$ e. $600 + 30 + 5 =$ ____

c. $80 + 2 =$ ____   f. $300 + 0 + 0 =$ ____

Lesson

Part 1

a.
$$K \xrightarrow{\quad J \quad} 74$$
$\boxed{J = 51}$
_____

b.
$$P \xrightarrow{\quad 24 \quad} M$$
$\boxed{P = 21}$
_____

c.
$$T \xrightarrow{\quad 64 \quad} B$$
$\boxed{B = 86}$
_____

Part 2

a. ▢ + ▢ = 38

new ▢ + ▢ = 38

b. ▢ + ▢ = 71

new ▢ + ▢ = 71

Part 3

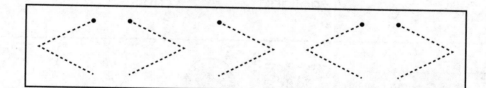

a. $\boxed{\;<\;}$

b. $\boxed{\;>\;}$

Part 4

a.
$$\begin{array}{r} \square \\ 5\,6 \\ +\,3\,2 \\ \hline \end{array}$$

b.
$$\begin{array}{r} \square \\ 3\,4 \\ +\,2\,9 \\ \hline \end{array}$$

c.
$$\begin{array}{r} \square \\ 7\,8 \\ +\,2\,1 \\ \hline \end{array}$$

d.
$$\begin{array}{r} \square \\ 6\,9 \\ +\,1\,3 \\ \hline \end{array}$$

Lesson 28

Name _____

Part 5

a. F is 12 more than B.
B is 77.
What number is F?

_____

b. P is 22 less than T.
P is 53.
What number is T?

_____

c. J is 73 more than M.
J is 85.
What number is M?

_____

Part 6

a.

_____ dollars   _____ cents

b.

_____ dollars   _____ cents

c.

_____ dollars   _____ cents

Connecting Math Concepts

Lesson 28

Name _____

Part 7

a. 20 – 10 = _____

b. 7 + 10 = _____

c. 16 – 6 = _____

d. 5 + 10 = _____

e. 14 – 4 = _____

f. 13 – 10 = _____

g. 12 – 2 = _____

h. 18 – 10 = _____

i. 9 + 10 = _____

j. 11 – 10 = _____

k. 15 – 5 = _____

l. 8 + 10 = _____

m. 4 + 10 = _____

n. 12 – 10 = _____

o. 11 – 1 = _____

p. 6 + 10 = _____

q. 10 + 7 = _____

r. 15 – 10 = _____

s. 16 – 10 = _____

t. 10 + 10 = _____

u. 17 – 7 = _____

v. 19 – 10 = _____

w. 18 – 8 = _____

x. 14 – 10 = _____

y. 3 + 10 = _____

z. 17 – 10 = _____

A. 19 – 9 = _____

B. 10 + 6 = _____

C. 1 + 10 = _____

D. 13 – 3 = _____

Independent Work

Part 8 Find the number for T.

a. 15 is 12 more than T.

_____→

b. T is 7 less than 18.

_____→ _____

c. T is 10 more than 36.

_____→

Lesson 28

Name _____

Part 9 Write the number of cents.

a.   _____ cents

b.   _____ cents

Part 10 Write the 3 numbers from smallest to biggest.

a. 31   29   17

___ ___ ___

b. 58   59   49

___ ___ ___

c. 60   55   58

___ ___ ___

Part 11

a. 2 – 1 = ____   b. 2 + 7 = ____   c. 6 – 4 = ____   d. 18 + 1 = ____

e. 12 – 10 = ____   f. 6 – 5 = ____   g. 2 + 5 = ____   h. 5 + 2 = ____

Part 12

a. 186 + 10 = _____   b. 113 + 10 = _____   c. 10 + 48 = _____

d. 10 + 39 = _____   e. 150 + 10 = _____

Connecting Math Concepts

Lesson

Name _____

Part 1

$$3 \xrightarrow{\quad 3 \quad} 6 \qquad 4 \xrightarrow{\quad 3 \quad} \blacksquare \qquad 5 \xrightarrow{\quad 3 \quad} \blacksquare \qquad 6 \xrightarrow{\quad 3 \quad} \blacksquare$$

Part 2

a. ▢ + ▢ = 23   b. ▢ + ▢ = 76

new ▢ + ▢ = 23   new ▢ + ▢ = 76

Part 3

a. R is 20 more than P.
R is 88.
What number is P?

_____

→

b. J is 32 more than M.
M is 55.
What number is J?

_____

→

c. K is 30 less than T.
T is 75.
What number is K?

_____

→

Part 4

a. 19 ▢ 20   b. 33 ▢ 22   c. 40 ▢ 39   d. 1 ▢ 7

Part 5

a. 3 8
   + 1 2

b. 4 5
   + 3 9

c. 5 9
   + 2 3

Lesson

Name _____

Part 6

a. $3 \xrightarrow{3} 6$ b. $4 \xrightarrow{3} \blacksquare$ c. $5 \xrightarrow{3} \blacksquare$ d. $6 \xrightarrow{3} \blacksquare$

_____ _____ _____ _____

_____ _____ _____ _____

Part 7

a.

_____ dollars _____ cents

b.

_____ dollars _____ cents

c.

_____ dollars _____ cents

Independent Work

Part 8

a.
```
  12
  57
+ 30
```

b.
```
  98
- 72
```

c.
```
  41
  26
+ 32
```

d.
```
  473
- 252
```

e.
```
  170
  206
+ 623
```

Connecting Math Concepts

Lesson 29

Name _____

Part 9

a. T is 11 more than 58.

_____

b. 23 is 14 less than P.

_____

c. R is 12 less than 34.

_____

d. 67 is 26 more than P.

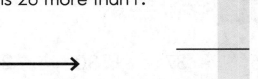

_____

Part 10 — Complete the place-value facts.

a. $500 + 90 + 0 =$ _____

b. $300 + 20 + 8 =$ _____

c. _____ $= 551$

d. _____ $= 83$

e. $200 + 0 + 0 =$ _____

f. _____ $= 400$

Part 11

a. $2 + 10 =$ ____

b. $17 - 10 =$ ____

c. $2 + 6 =$ ____

d. $8 - 7 =$ ____

e. $1 + 11 =$ ____

f. $9 - 7 =$ ____

g. $7 + 10 =$ ____

h. $2 + 9 =$ ____

i. $16 - 6 =$ ____

j. $9 + 2 =$ ____

k. $13 - 10 =$ ____

l. $8 - 6 =$ ____

m. $10 + 5 =$ ____

n. $4 + 10 =$ ____

o. $10 + 10 =$ ____

p. $9 - 8 =$ ____

Lesson

Name _____

Part 1

a. 3 → 3 → ■   b. 4 → 3 → ■   c. 5 → 3 → ■   d. 6 → 3 → ■

_____ _____ _____

_____ _____ _____

Part 2

a. 36   37   b. 30   20   c. 59   60   d. 30   29

Part 3

a. T is 18 less than B.
B is 79.
What number is T?

_____ →

b. R is 54 less than M.
R is 31.
What number is M?

_____ →

c. W is 22 less than P.
W is 76.
What number is P?

_____ →

d. Y is 13 more than J.
Y is 95.
What number is J?

_____ →

Part 4

a. 6
− 4

b. 6
− 8

c. 6
− 9

d. 6
− 6

e. 6
− 7

Part 5

a. 7 6
+ 1 9

b. 6 9
+ 2 3

c. 2 9
+ 5 2

d. 4 8
+ 3 3

Lesson 30

Name _____

Part 6

↓
_____ ☐ inches

| | 1| | 2| | 3| | 4| | 5| | 6| |

Independent Work

Part 7 Write the simple place-value fact. Below, write the new place-value fact.

a. ☐ + ☐ = 62

new ☐ + ☐ = 62

b. ☐ + ☐ = 78

new ☐ + ☐ = 78

c. ☐ + ☐ = 44

new ☐ + ☐ = 44

Part 8 Write the dollars and cents for each row.

a.

☐ dollars ☐ cents

b.

☐ dollars ☐ cents

c.

☐ dollars ☐ cents

Connecting Math Concepts

Lesson 30

Name _____

Copyright © The McGraw-Hill Companies, Inc.

Part 9

a. $\begin{array}{r} 6 \\ +\ 10 \\ \hline \end{array}$
b. $\begin{array}{r} 6 \\ +\ 9 \\ \hline \end{array}$
c. $\begin{array}{r} 7 \\ +\ 10 \\ \hline \end{array}$
d. $\begin{array}{r} 7 \\ +\ 9 \\ \hline \end{array}$
e. $\begin{array}{r} 5 \\ +\ 10 \\ \hline \end{array}$
f. $\begin{array}{r} 5 \\ +\ 9 \\ \hline \end{array}$

Part 10

a. 40 is 16 less than J.

b. P is 36 more than 12.

_____ _____

Part 11 Write the missing numbers.

a. __45__ __50__ _____ _____ _____ _____ __75__ _____

Part 12

a. $16 - 10 =$ ____

b. $8 - 6 =$ ____

c. $16 - 6 =$ ____

d. $2 + 5 =$ ____

e. $8 + 10 =$ ____

f. $16 - 6 =$ ____

g. $8 + 2 =$ ____

h. $5 + 10 =$ ____

i. $10 + 6 =$ ____

j. $9 - 8 =$ ____

k. $1 + 8 =$ ____

l. $18 - 8 =$ ____

m. $16 - 1 =$ ____

n. $3 + 10 =$ ____

o. $10 + 9 =$ ____

p. $9 - 2 =$ ____

Connecting Math Concepts

Lesson 31

Name _____

Part 1

a. 3 ——3——▶■   b. 4 ——3——▶■   c. 5 ——3——▶■   d. 6 ——3——▶■

_____ _____ _____ _____

_____ _____ _____ _____

Part 2

a. 7   b. 4   c. 6   d. 10   e. 12   f. 8   g. 7
  − 8   − 6   − 4   − 10   − 13   − 8   − 6

Part 3

a. M is 56 less than J.
   M is 30.
   What number is J?   _____

   ——————————▶

b. R is 20 less than P.
   P is 85.
   What number is R?   _____

   ——————————▶

c. T is 12 more than V.
   T is 54.
   What number is V?   _____

   ——————————▶

d. M is 72 more than F.
   F is 23.
   What number is M?   _____

   ——————————▶

Part 4

a. 11   10   b. 89   90   c. 52   60

d. 80   70   e. 15   14   f. 100   200

Lesson 31

Name _____

Part 5

a. →

b. →

c. →

Part 6

a.
$$\begin{array}{r} 3\,5 \\ +\,2\,3 \\ \hline \end{array}$$

b.
$$\begin{array}{r} 7\,8 \\ +\,1\,9 \\ \hline \end{array}$$

c.
$$\begin{array}{r} 4\,5 \\ +\,3\,9 \\ \hline \end{array}$$

d.
$$\begin{array}{r} 2\,8 \\ +\,3\,1 \\ \hline \end{array}$$

Part 7

a. _____ ▢ inches

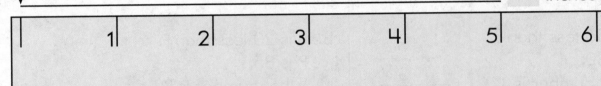

| | 1 | 2 | 3 | 4 | 5 | 6 |

b. _____ ▢ inches

c. _____ ▢ inches

Independent Work

Part 8 Write the simple place-value fact. Below, write the new place-value fact.

a. ▢ + ▢ = 62

new ▢ + ▢ = 62

b. ▢ + ▢ = 78

new ▢ + ▢ = 78

c. ▢ + ▢ = 44

new ▢ + ▢ = 44

Connecting Math Concepts

Lesson

Name _____

Part 9 Write the missing numbers.

a. _____ _____ 27 36 _____ _____ _____ _____ _____

b. 16 18 _____ _____ _____ _____ 26 _____

Part 10

a. 426
  + 303

b. 489
  − 360

c. 765
  + 123

d. 985
  − 863

Part 11

a. 44 is 31 more than K.

b. 88 is 11 less than P.

Part 12 Complete the place-value facts.

a. _____ = 309   b. 700 + 70 + 3 = _____

Part 13 Write the dollars and cents for the row.

_____ dollars _____ cents

Lesson

Name _____

Part 1

a. J is 200 less than M.
    J is 680.
    What number is M? _____

b. T is 19 less than R.
    R is 549.
    What number is T? _____

c. P is 201 more than H.
    P is 322.
    What number is H? _____

d. Y is 136 more than W.
    W is 502.
    What number is Y? _____

Part 2

a. $\begin{array}{r} 7 \\ -\ 5 \\ \hline \end{array}$
b. $\begin{array}{r} 4 \\ -\ 3 \\ \hline \end{array}$
c. $\begin{array}{r} 8 \\ -\ 9 \\ \hline \end{array}$
d. $\begin{array}{r} 7 \\ -\ 6 \\ \hline \end{array}$
e. $\begin{array}{r} 8 \\ -\ 8 \\ \hline \end{array}$
f. $\begin{array}{r} 5 \\ -\ 6 \\ \hline \end{array}$

Part 3

a. 7 6

b. 4 4

c. 11 12

d. 50 60

e. 9 8

f. 13 13

Part 4

a. ⟶

b. ⟶

c. ⟶

d. ⟶

Connecting Math Concepts

Lesson 32

Part 5

a. ↓ _____ ⬜ inches

b. ↓ _____ ⬜ inches

c. ↓ _____ ⬜ inches

Part 6

a. 6 ——3——▶ ■

b. 7 ——3——▶ ■

c. 8 ——3——▶ ■

_____ _____ _____

_____ _____ _____

Independent Work

Part 7

a. 4 8
   + 3 3
   _____

b. 8 6
   + 1 2
   _____

c. 5 7
   + 2 9
   _____

d. 1 5
   + 8 1
   _____

Part 8

a. P is 145 less than 366.

————▶   _____ ⬜ ⬜

b. H is 270 more than 625.

————▶   _____ ⬜ ⬜

Part 9

a. 3 0 7
   − 1 0 5
   _____

b. 5 2 6
   + 3 3 3
   _____

c.   4 6
     3 1 2
   + 2 0
   _____

d. 8 9 7
   − 2 7 7
   _____

Lesson 32

Name _____

Part 10

a. $15 - 10 =$ _____   g. $19 - 9 =$ _____

b. $16 - 6 =$ _____   h. $2 + 7 =$ _____

c. $13 - 12 =$ _____   i. $2 + 8 =$ _____

d. $20 - 10 =$ _____   j. $5 - 3 =$ _____

e. $8 + 10 =$ _____   k. $11 - 2 =$ _____

f. $1 + 12 =$ _____   l. $4 + 2 =$ _____

Part 11 Write the 3 numbers from smallest to biggest.

	88	86	85
a.	_____	_____	_____

	223	217	219
b.	_____	_____	_____

	786	785	784
c.	_____	_____	_____

Part 12 Write the dollars and cents for each row.

a.

_____ dollars _____ cents

b.

_____ dollars _____ cents

Part 13 Write the simple place-value fact. Below write the new place-value fact.

a. _____ + _____ $= 40$ b. _____ + _____ $= 81$

new _____ + _____ $= 40$ new _____ + _____ $= 81$

c. _____ + _____ $= 26$ d. _____ + _____ $= 99$

new _____ + _____ $= 26$ new _____ + _____ $= 99$

Connecting Math Concepts

Lesson

Name _____

Part 1

a. 5 x 3 = _____   b. 10 x 4 = _____

Part 2

a. ↓ _____ ▢ inches

b. ↓ _____ ▢ inches

Part 3

a. Fran made 9 more pictures.   b. Bob lost 14 dollars.

→   →

Part 4

a. 3 ——→ 3 → ■   b. 4 ——→ 3 → ■   c. 5 ——→ 3 → ■

_____   _____   _____

_____   _____   _____

d. 6 ——→ 3 → ■   e. 7 ——→ 3 → ■   f. 8 ——→ 3 → ■

_____   _____   _____

_____   _____   _____

Lesson 33

Name _____

Part 5 Figure out the missing number.

a. B is 10 less than J.
B is 61.
What number is J? _____

b. R is 22 more than P.
P is 50.
What number is R? _____

c. V is 20 less than X.
X is 90.
What number is V? _____

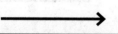

Part 6

a.
```
   8 1 0
   1 2 5
 +   3 3
```

b.
```
   5 9 6
 − 2 7 1
```

c.
```
   4 2 9
 + 3 2 2
```

d.
```
   3 7
 + 2 3
```

Part 7 Write the dollars and cents for each row.

a.

_____ dollars _____ cents

b.

_____ dollars _____ cents

Connecting Math Concepts

Lesson 33

Name _____

Part 8 Write an addtion and a subtraction fact for each family.

a. $\xrightarrow{\quad 3 \qquad 3 \quad}$ ■

_____

_____

b. $\xrightarrow{\quad 4 \qquad 3 \quad}$ ■

_____

_____

c. $\xrightarrow{\quad 5 \qquad 3 \quad}$ ■

_____

_____

Part 9

a. $14 - 4 =$ _____

b. $18 - 10 =$ _____

c. $12 - 10 =$ _____

d. $17 - 7 =$ _____

e. $4 + 10 =$ _____

f. $2 + 5 =$ _____

g. $2 + 9 =$ _____

h. $9 - 7 =$ _____

i. $8 - 2 =$ _____

j. $12 - 11 =$ _____

k. $6 - 4 =$ _____

l. $1 + 15 =$ _____

Part 10 Write the simple place-value fact. Below write the new place-value fact.

a. ▢ + ▢ $= 87$

new ▢ + ▢ $= 87$

b. ▢ + ▢ $= 50$

new ▢ + ▢ $= 50$

c. ▢ + ▢ $= 65$

new ▢ + ▢ $= 65$

Lesson Name _____

Part 1

a. 2 x 5 = _____   b. 1 x 6 = _____   c. 5 x 4 = _____

Part 2

a. Ms. Jones gave away 12 cups.

⟶

b. Ann sold 2 pictures.

⟶

c. Jim earned 76 dollars.

⟶

d. Lynn lost 3 pictures.

⟶

e. Mr. Jones found 23 coins.

⟶

Part 3

a. 16 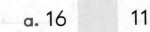 11

b. 12 ▢ 21

c. J  E

Part 4

a. ↓ _____ ▢ centimeters

b. ↓ _____ ▢ inches

Part 5

a. 3 ⟶ 3 ■

_____

b. 4 ⟶ 3 ■

_____

_____

c. 5 ⟶ 3 ■

_____

_____

d. 6 ⟶ 3 ■

_____

e. 7 ⟶ 3 ■

_____

f. 8 ⟶ 3 ■

_____

Connecting Math Concepts

Lesson 34

Name _____

Independent Work

Part 6 Cross out problems you cannot work. Work the rest of the problems.

a. $\begin{array}{r} 6 \\ -3 \\ \hline \end{array}$ b. $\begin{array}{r} 1 \\ -0 \\ \hline \end{array}$ c. $\begin{array}{r} 6 \\ -7 \\ \hline \end{array}$ d. $\begin{array}{r} 0 \\ -5 \\ \hline \end{array}$ e. $\begin{array}{r} 5 \\ -0 \\ \hline \end{array}$ f. $\begin{array}{r} 4 \\ -5 \\ \hline \end{array}$

Part 7

a. T is 11 more than J.
T is 56.
What number is J? _____

b. K is 17 less than P.
P is 58.
What number is K? _____

c. R is 12 less than Y.
R is 87.
What number is Y? _____

Part 8

a. $10 + 81 =$ _____   b. $10 + 653 =$ _____   c. $403 + 10 =$ _____

d. $260 + 10 =$ _____   e. $13 + 10 =$ _____

Part 9 Write an addition and a subtraction fact for each family.

a. $\xrightarrow{\hspace{0.3cm}3\hspace{1cm}3\hspace{0.3cm}}$ ■

b. $\xrightarrow{\hspace{0.3cm}4\hspace{1cm}3\hspace{0.3cm}}$ ■

c. $\xrightarrow{\hspace{0.3cm}5\hspace{1cm}3\hspace{0.3cm}}$ ■

_____   _____   _____

_____   _____   _____

Lesson 34

Name _____

Part 10

a. $4 + 10 =$ _____

b. $2 + 9 =$ _____

c. $1 + 18 =$ _____

d. $7 + 10 =$ _____

e. $2 + 8 =$ _____

f. $10 + 6 =$ _____

g. $1 + 7 =$ _____

h. $3 + 10 =$ _____

i. $5 + 2 =$ _____

j. $10 + 2 =$ _____

k. $8 + 10 =$ _____

l. $2 + 9 =$ _____

m. $1 + 8 =$ _____

n. $8 + 2 =$ _____

o. $10 + 9 =$ _____

Part 11 Write the place-value facts.

a. ☐ + ☐ = 60

new ☐ + ☐ = 60

b. ☐ + ☐ = 28

new ☐ + ☐ = 28

c. ☐ + ☐ = 36

new ☐ + ☐ = 36

Part 12

a. $\begin{array}{r} 48 \\ +22 \\ \hline \end{array}$

b. $\begin{array}{r} 56 \\ +19 \\ \hline \end{array}$

c. $\begin{array}{r} 15 \\ +39 \\ \hline \end{array}$

d. $\begin{array}{r} 28 \\ +63 \\ \hline \end{array}$

Connecting Math Concepts

Lesson

Name _____

Part 1

a. 10 x 6 = _____   b. 5 x 3 = _____   c. 9 x 7 = _____

Part 2

a. Andy sold 17 bikes.

b. Dan gave away 25 dollars.

c. Tina earned 200 dollars.

d. They bought 19 books.

e. Her mother lost 21 pounds.

Part 3

a. 20 [ ] 22

b. 80 [ ] 60

c. P [ ] M

Part 4 Show the new place-value for each number.

a. 5 3

b. 7 4

Part 5

a. ↓ _____ [ ] inches

b. ↓ _____ [ ] centimeters

c. ↓ _____ [ ] inches

d. ↓ _____ [ ] centimeters

Lesson 35

Name _____

Part 6

a. 4 ―――3→ ■

_____

_____

b. 8 ―――3→ ■

_____

_____

c. 6 ―――3→ ■

_____

_____

d. 3 ―――3→ ■

_____

_____

e. 7 ―――3→ ■

_____

_____

_____

f. 5 ―――3→ ■

_____

_____

Independent Work

Part 7 Cross out problems you cannot work. Work the rest of the problems.

a. 6
   − 8

b. 3
   − 1

c. 5
   − 7

d. 4
   − 3

e. 8
   − 6

f. 9
   − 8

Part 8

a. 4 1 8
  + 2 5 2

b. 4 2 8
    3 0 1
  + 2 1 3

c. 4 9
  + 2 3

Part 9 Write an addition and a subtraction fact for each family.

a. 3 ―――3→ ■

_____

_____

b. 4 ―――3→ ■

_____

_____

c. 5 ―――3→ ■

_____

_____

Connecting Math Concepts

Lesson 35

Name _____

Part 10

a. 16 – 10 = _____   g. 9 – 7 = _____   m. 8 – 2 = _____

b. 5 – 4 = _____   h. 11 – 9 = _____   n. 19 – 9 = _____

c. 17 – 7 = _____   i. 7 – 6 = _____   o. 5 – 3 = _____

d. 12 – 2 = _____   j. 18 – 8 = _____   p. 7 – 6 = _____

e. 10 – 8 = _____   k. 4 – 3 = _____   q. 13 – 3 = _____

f. 7 – 1 = _____   l. 11 – 9 = _____   r. 11 – 1 = _____

Part 11

a. ____ + ____ = 46

new ____ + ____ = 46

b. ____ + ____ = 79

new ____ + ____ = 79

c. ____ + ____ = 30

new ____ + ____ = 30

Part 12

a. V is 16 less than J.
J is 88.
What number is V?

_____ ⟶

b. P is 72 more than W.
P is 99.
What number is W?

_____ ⟶

c. T is 54 more than J.
J is 12.
What number is T?

_____ ⟶

Lesson

Name _____

Part 1

a. 2 x 6 = _____   b. 10 x 3 = _____   c. 5 x 7 = _____

Part 2

a. 4 3 + 2   b. 10 – 1 [ ] 7

c. 9 + 1 [ ] 12   d. 5 [ ] 9 – 2

Part 3

a. 8 3   b. 2 5   c. 5 8

Part 4

a.   b.   c. inches

Part 5

a. 8 ——3→ ■   b. 5 ——3→ ■   c. 4 ——3→ ■

_____   _____   _____

_____   _____   _____

d. 7 ——3→ ■   e. 6 ——3→ ■   f. 3 ——3→ ■

_____   _____   _____

_____   _____   _____

Connecting Math Concepts

Lesson 36

Name _____

Independent Work

Part 6 Write the sign > or <.

a. 56 ▢ 58   b. 734 ▢ 534   c. 200 ▢ 100   d. 204 ▢ 206

Part 7 Write the 3 numbers from smallest to biggest.

| 156 | 145 | 155 |

a. _____ _____ _____

| 320 | 317 | 322 |

b. _____ _____ _____

| 786 | 785 | 784 |

c. _____ _____ _____

Part 8 Write the place-value facts.

a. ▢ + ▢ = 22

new ▢ + ▢ = 22

b. ▢ + ▢ = 53

new ▢ + ▢ = 53

c. ▢ + ▢ = 91

new ▢ + ▢ = 91

Part 9

a. 3 1 8
  + 1 0 3

b. 5 9
  + 3 2

c. 1 8
    6 2
  + 1 0

Part 10 Complete the place-value facts.

a. 30 + 9 = _____

b. 500 + 0 + 9 = _____

c. _____ = 139

d. 700 + 0 + 0 = _____

Connecting Math Concepts

Lesson 36 **91**

Lesson 36

Name _____

Part 11 | Write the dollars and cents for each row.

a.

[   ] dollars   [   ] cents

b.

[   ] dollars   [   ] cents

Part 12

a. J is 80 less than P.
    P is 99.
    What number is J?   _____

b. R is 75 more than T.
    R is 87.
    What number is T?   _____

c. K is 73 more than F.
    F is 21.
    What number is K?   _____

Part 13

a. 71 + 10 = _____   b. 10 + 65 = _____   c. 126 + 10 = _____

d. 10 + 165 = _____   e. 10 + 148 = _____   f. 170 + 10 = _____

Lesson Name _____

Part 1

a. $9 \times 5 =$ _____

b. $2 \times 6 =$ _____

Part 2

a. There were 38 people on the bus.
Then 12 people got off the bus.
How many people ended up on
the bus?

→

b. There were 25 people on a plane.
Then 42 people got on the plane.
How many people ended up on
the plane?

→

c. Tom started with 18 corn chips.
He ate 10 chips.
How many chips did he still have?

→

Part 3

____ 8 ____ 16 ____

____ 28 ____ 36 ____

Part 4

a. $5 + 3$ ▢ 10

b. 6 ▢ $7 - 2$

c. $4 + 3$ ▢ 8

d. $12 - 2$ ▢ 9

Part 5

a. 3 6　　　b. 7 4　　　c. 2 7

Part 6

inches

a. ▢ 　　 b. ▢ 　 c. ▢ 　 end

Lesson 37

Name _____

a. 3 + 7 = _____   j. 10 − 3 = _____   s. 3 + 7 = _____

b. 3 + 4 = _____   k. 6 − 3 = _____   t. 3 + 8 = _____

c. 3 + 8 = _____   l. 8 + 3 = _____   u. 9 − 3 = _____

d. 6 + 3 = _____   m. 9 − 3 = _____   v. 8 − 5 = _____

e. 11 − 3 = _____   n. 3 + 6 = _____   w. 3 + 4 = _____

f. 9 − 3 = _____   o. 12 − 9 = _____   x. 11 − 3 = _____

g. 7 + 3 = _____   p. 10 − 3 = _____   y. 10 − 3 = _____

h. 11 − 8 = _____   q. 7 − 3 = _____   z. 3 + 6 = _____

i. 7 − 3 = _____   r. 7 + 3 = _____

Independent Work

Part 8 Write 2 addition facts.

a. 4 ———— 3 → ■

_____

_____

b. 5 ———— 3 → ■

_____

_____

c. 6 ———— 3 → ■

_____

_____

Part 9

a. 8   b. 7   c. 8   d. 6   e. 7   f. 9
  − 3   − 3   − 5   − 3   − 4   − 6

Lesson 37

Name _____

Part 10 Write the place-value facts.

a. ▢ + ▢ = 87

new ▢ + ▢ = 87

b. ▢ + ▢ = 60

new ▢ + ▢ = 60

c. ▢ + ▢ = 32

new ▢ + ▢ = 32

Part 11

a. $19 - 10 =$ _____

b. $5 - 1 =$ _____

c. $9 - 2 =$ _____

d. $5 - 2 =$ _____

e. $15 - 5 =$ _____

f. $9 - 8 =$ _____

g. $15 - 10 =$ _____

h. $11 - 9 =$ _____

i. $8 - 2 =$ _____

j. $5 - 4 =$ _____

k. $3 - 1 =$ _____

l. $12 - 10 =$ _____

m. $19 - 9 =$ _____

n. $7 - 2 =$ _____

o. $12 - 10 =$ _____

p. $5 - 1 =$ _____

Part 12

a.
```
  4 6 3
  2 1 1
+ 3 2 3
```

b.
```
  5 8 6
- 4 7 2
```

c.
```
  6 1 5
+ 1 7 9
```

Part 13 Complete the place-value facts.

a. _____ $= 67$

b. $700 + 10 + 1 =$ _____

Lesson 37

Name _____

Part 14

a. $204 + 10 =$ _____

b. $10 + 107 =$ _____

c. $58 + 10 =$ _____

d. $780 + 10 =$ _____

e. $10 + 512 =$ _____

f. $10 + 188 =$ _____

Part 15

a. T is 41 more than R.
   R is 8.
   What number is T?

_____

b. P is 77 less than T.
   T is 88.
   What number is P?

_____

c. V is 36 less than M.
   V is 12.
   What number is M?

_____

Part 16

a. $\begin{array}{r} 8 \\ +\ 10 \\ \hline \end{array}$

b. $\begin{array}{r} 8 \\ +\ 9 \\ \hline \end{array}$

c. $\begin{array}{r} 7 \\ +\ 10 \\ \hline \end{array}$

d. $\begin{array}{r} 7 \\ +\ 9 \\ \hline \end{array}$

e. $\begin{array}{r} 5 \\ +\ 10 \\ \hline \end{array}$

f. $\begin{array}{r} 5 \\ +\ 9 \\ \hline \end{array}$

g. $\begin{array}{r} 4 \\ +\ 10 \\ \hline \end{array}$

h. $\begin{array}{r} 4 \\ +\ 9 \\ \hline \end{array}$

i. $\begin{array}{r} 9 \\ +\ 10 \\ \hline \end{array}$

j. $\begin{array}{r} 9 \\ +\ 9 \\ \hline \end{array}$

k. $\begin{array}{r} 6 \\ +\ 10 \\ \hline \end{array}$

l. $\begin{array}{r} 6 \\ +\ 9 \\ \hline \end{array}$

Lesson

Name _____

Part 1

a. 10 x 5 = _____   b. 2 x 8 = _____   c. 5 x 7 = _____

Part 2

a. Bill had 25 toys.
   Then he made 11 toys.
   How many toys did Bill end up with?

   ⟶

b. Don had 58 books.
   He lost 18 books.
   How many books did he end up with?

   ⟶

c. Dolly had 8 bags.
   She bought 30 more bags.
   How many bags did she end up with?

   ⟶

Part 3

4 ___ ___ ___ 16 ___

___ ___ 32 ___ ___

Part 4

a. 4 – 2 ▢ 3

b. 10 ▢ 7 + 3

c. 14 ▢ 10 + 3

d. 9 + 2 ▢ 11

e. 7 – 2 ▢ 4

Part 5

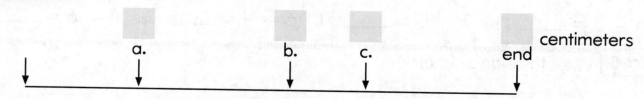

a.   b. c.   end centimeters

Lesson 38

Name _____

Part 6

a. $3 + 8 =$ _____

b. $4 + 3 =$ _____

c. $10 - 3 =$ _____

d. $7 - 3 =$ _____

e. $3 + 6 =$ _____

f. $11 - 3 =$ _____

g. $3 + 7 =$ _____

h. $9 - 3 =$ _____

i. $11 - 8 =$ _____

j. $9 + 3 =$ _____

k. $12 - 9 =$ _____

l. $7 - 3 =$ _____

m. $10 - 3 =$ _____

n. $3 + 6 =$ _____

o. $7 - 4 =$ _____

p. $5 + 3 =$ _____

q. $3 + 7 =$ _____

r. $8 + 3 =$ _____

s. $9 - 3 =$ _____

t. $10 - 7 =$ _____

u. $3 + 4 =$ _____

v. $6 + 3 =$ _____

w. $11 - 3 =$ _____

x. $5 - 2 =$ _____

y. $3 + 7 =$ _____

z. $9 - 3 =$ _____

Independent Work

Part 7

a. $8 - 2 =$ _____	e. $12 - 2 =$ _____	i. $6 - 2 =$ _____	m. $18 - 10 =$ _____
b. $11 - 9 =$ _____	f. $8 - 7 =$ _____	j. $12 - 2 =$ _____	n. $7 - 2 =$ _____
c. $7 - 6 =$ _____	g. $15 - 5 =$ _____	k. $13 - 3 =$ _____	o. $11 - 9 =$ _____
d. $5 - 3 =$ _____	h. $7 - 5 =$ _____	l. $6 - 4 =$ _____	p. $8 - 6 =$ _____

Part 8 Write the sign >, <, or =.

a. 56 ☐ 36   b. 21 ☐ 18   c. 38 ☐ 38   d. 51 ☐ 61

Connecting Math Concepts

Lesson 38

Name _____

Part 9 Write the place-value facts.

a. ☐ + ☐ = 94

new ☐ + ☐ = 94

b. ☐ + ☐ = 75

new ☐ + ☐ = 75

c. ☐ + ☐ = 20

new ☐ + ☐ = 20

Part 10 Write the missing numbers.

__20__ __25__ ____ ____ ____ ____ ____ ____ __60__

Part 11 Cross out problems you cannot work. Work the rest of the problems.

a. 7
   − 9

b. 7
   − 0

c. 3
   − 2

d. 3
   − 8

e. 9
   − 7

Part 12

a. W is 30 more than P.
   P is 26.
   What number is W? _____

b. K is 47 more than J.
   K is 68.
   What number is J? _____

Part 13

a. 9
   − 3

b. 7
   − 4

c. 7
   − 3

d. 9
   − 6

e. 5
   − 2

f. 6
   − 3

g. 5
   − 3

Part 14

a. 26 + 10 = _____

b. 107 + 10 = _____

c. 10 + 56 = _____

d. 483 + 10 = _____

e. 10 + 103 = _____

f. 10 + 920 = _____

Lesson

Name _____

Part 1

a. $9 \times 6 =$ _____   b. $1 \times 4 =$ _____

c. $10 \times 5 =$ _____   d. $4 \times 3 =$ _____

Part 2

____ 4 ____ ____ ____ ____

____ ____ ____ 32 ____ ____

Part 3

a. 23 ☐ $20 + 3$ b. $10 + 4$ ☐ 11   c. $5 + 3$ ☐ 8

d. 10 ☐ $11 - 2$ e. $6 - 1$ ☐ 7

Part 4

a. 74   b. 90   c. 37   d. 82

Part 5

cm ☐   in. ☐   in.   cm ☐

a.   b.   c.   end

Connecting Math Concepts

Lesson 39

Name _____

Part 6

a. $3 + 3 =$ ____

b. $9 - 3 =$ ____

c. $3 + 5 =$ ____

d. $6 - 3 =$ ____

e. $8 - 3 =$ ____

f. $10 - 3 =$ ____

g. $3 + 6 =$ ____

h. $11 - 3 =$ ____

i. $6 - 3 =$ ____

j. $4 + 3 =$ ____

k. $3 + 3 =$ ____

l. $5 + 3 =$ ____

m. $8 + 3 =$ ____

n. $3 + 3 =$ ____

o. $6 - 3 =$ ____

p. $10 - 7 =$ ____

q. $7 - 3 =$ ____

r. $6 + 3 =$ ____

s. $11 - 8 =$ ____

t. $3 + 5 =$ ____

u. $9 - 3 =$ ____

v. $12 - 9 =$ ____

w. $8 - 3 =$ ____

x. $3 + 4 =$ ____

Independent Work

Part 7 Write the missing numbers.

a. __14__ __16__ __18__ ____ ____ ____ ____ ____

b. __36__ __45__ __54__ ____ ____ ____ ____

Part 8

a. $\begin{array}{r} 16 \\ -\ 6 \\ \hline \end{array}$

b. $\begin{array}{r} 9 \\ -\ 2 \\ \hline \end{array}$

c. $\begin{array}{r} 12 \\ -\ 2 \\ \hline \end{array}$

d. $\begin{array}{r} 9 \\ -\ 7 \\ \hline \end{array}$

e. $\begin{array}{r} 3 \\ +\ 2 \\ \hline \end{array}$

f. $9 + 1 =$ ____

g. $2 + 6 =$ ____

h. $5 + 10 =$ ____

i. $6 - 4 =$ ____

j. $3 - 2 =$ ____

k. $2 + 9 =$ ____

Copyright © The McGraw-Hill Companies, Inc.

Connecting Math Concepts

Lesson 39 **101**

Lesson 39

Name _____

Part 9 Write 2 subtraction facts.

a. ■ ———$\xrightarrow{3}$ 9

_____

_____

b. ■ ———$\xrightarrow{3}$ 7

_____

_____

c. ■ ———$\xrightarrow{3}$ 10

_____

_____

d. ■ ———$\xrightarrow{3}$ 6

_____

_____

e. ■ ———$\xrightarrow{3}$ 8

_____

_____

f. ■ ———$\xrightarrow{3}$ 5

_____

_____

Part 10

_____ dollars   _____ cents

Part 11

a. K is 12 more than B.
B is 24.
What number is K?   _____

b. 56 is 11 less than R.
What number is R?   _____

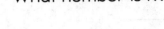

c. J is 20 less than P.
J is 38.
What number is P?   _____

Part 12

a. 5 8 7
   − 8 2

b. 4 1 8
   + 3 6 3

c. 4 8 6
   − 1 6 6

Connecting Math Concepts

Lesson

Name _____

Part 1

a. 91   _____   b. 37   _____

Part 2

          cm   in.   cm   in.

          a.   b.   c.   end

Part 3

a. 5 x 4 = _____   b. 2 x 8 = _____

c. 4 x 5 = _____   d. 9 x 3 = _____

Part 4

a. 41   b. 80   c. 65   d. 38

Part 5

a. _____ cents   _____ cents

b. _____ cents   _____ cents

c. _____ cents   _____ cents

Lesson 40

Part 6

__4__ ___ ___ ___ ___

___ ___ ___ ___ ___

Part 7

a. $3 + 5 =$ ___

b. $3 + 3 =$ ___

c. $10 - 3 =$ ___

d. $8 + 3 =$ ___

e. $6 + 3 =$ ___

f. $7 - 3 =$ ___

g. $8 - 3 =$ ___

h. $10 - 7 =$ ___

i. $6 - 3 =$ ___

j. $3 + 8 =$ ___

k. $3 + 6 =$ ___

l. $3 + 4 =$ ___

m. $9 - 6 =$ ___

n. $7 + 3 =$ ___

o. $8 - 3 =$ ___

p. $3 + 5 =$ ___

q. $8 + 3 =$ ___

r. $3 + 3 =$ ___

s. $11 - 3 =$ ___

t. $6 - 3 =$ ___

u. $4 + 3 =$ ___

v. $8 + 3 =$ ___

w. $8 - 5 =$ ___

x. $10 - 3 =$ ___

y. $3 + 5 =$ ___

z. $8 - 3 =$ ___

A. $3 + 7 =$ ___

B. $3 + 3 =$ ___

C. $6 - 3 =$ ___

D. $9 + 3 =$ ___

Connecting Math Concepts

Lesson 40

Name _____

Part 8

a. Bill started out with some marbles.
Then he gave 10 marbles away.
He ended up with 47 marbles.
How many marbles did he start out with?

————

⟶

b. There were some passengers on a bus.
Then 11 passengers got on the bus.
The bus ended up with 14 passengers.
How many passengers started out on the bus?

————

⟶

c. John had some potato chips.
He ate 14 chips.
He ended up with 3 chips.
How many chips did he start out with?

————

⟶

Independent Work

Part 9 | Write the sign >, <, or =.

a. 35 ▨ 30 + 5   b. 20 + 7 ▨ 26

c. 40 ▨ 45 – 1   d. 15 – 5 ▨ 11

Part 10

a. 804 + 10 = _____   b. 461 + 10 = _____

c. 23 + 10 = _____   d. 88 + 10 = _____

Lesson 40

Name _____

Part 11 Write the missing numbers.

a. __74__ __76__ ____ ____ ____

b. __65__ __70__ ____ ____ ____ ____

Part 12

a. 18   b. 4   c. 12   d. 9   e. 6   f. 1   g. 8   h. 5
  − 10   + 2   − 2   − 7   − 4   + 5   − 6   − 4

Part 13

a. 4 7 6   b. 5 3 2   c. 7 1 8
  − 3 5 4   + 1 3 8   − 5 1 6

Part 14 Write 2 subtraction facts.

a. ■ ───3──▶ 9

_____

_____

b. ■ ───3──▶ 8

_____

_____

c. ■ ───3──▶ 6

_____

_____

d. ■ ───3──▶ 10

_____

_____

e. ■ ───3──▶ 11

_____

_____

f. ■ ───3──▶ 5

_____

_____

Connecting Math Concepts

Lesson 41

Name _____

Part 1

a. _____ cents   _____ cents

b. _____ cents   _____ cents

c. _____ cents   _____ cents

d. _____ cents   _____ cents

Part 2

in.	cm		in.	cm
a.	b.		c.	end

Part 3

a. 3   b. 6   c. 9   d. 8   e. 7   f. 3   g. 10   h. 7
  + 4   − 3   − 3   + 3   − 3   + 6   − 3   + 3

i. 3   j. 7   k. 8   l. 3   m. 11   n. 5   o. 9   p. 3
  + 5   − 4   − 3   + 3   − 3   + 3   − 3   + 4

q. 11   r. 3   s. 3   t. 10   u. 8   v. 3   w. 6   x. 6
  − 3   + 3   + 8   − 7   − 3   + 7   + 3   − 3

Copyright © The McGraw-Hill Companies, Inc.

Lesson 41

Name _____

Part 4

a. A bus started out with some people.
   Then 12 more people got on the bus.
   35 people ended up on the bus.
   How many people did the bus start out with? 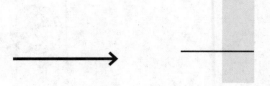 _____

b. A truck started out with some boxes on it.
   Then the driver took 67 boxes off the truck.
   The truck ended up with 21 boxes on it.
   How many boxes did the truck start out with? ⟶ _____

Independent Work

Part 5

a. $\begin{array}{r} 10 \\ -\ 9 \\ \hline \end{array}$
b. $\begin{array}{r} 12 \\ -\ 3 \\ \hline \end{array}$
c. $\begin{array}{r} 10 \\ -\ 2 \\ \hline \end{array}$
d. $\begin{array}{r} 11 \\ -\ 2 \\ \hline \end{array}$
e. $\begin{array}{r} 2 \\ +\ 6 \\ \hline \end{array}$
f. $\begin{array}{r} 14 \\ -\ 10 \\ \hline \end{array}$
g. $\begin{array}{r} 2 \\ +\ 8 \\ \hline \end{array}$
h. $\begin{array}{r} 5 \\ +\ 2 \\ \hline \end{array}$

Part 6 Write the missing numbers.

a.   _20_ _22_ ____ ____ _28_ ____

b.   _4_ _8_ ____ ____ ____

   ____ ____ ____ ____ _40_

Connecting Math Concepts

Lesson 41

Name _____

Part 7 Write the sign >, <, or =.

a. $5 + 3 + 2$  9

b. 17 $7 + 10$

c. 43 $40 + 4$

d. $20 + 5$ 25

Part 8

a. P is 30 less than J.
J is 51.
What number is P?

_____

⟶

b. 22 is 15 less than P.
What number is P?

_____

⟶

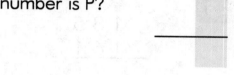

c. M is 16 more than K.
K is 67.
What number is M?

_____

⟶

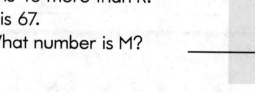

Part 9

a.
$$\begin{array}{r} 5\ 4\ 8 \\ +2\ 2\ 2 \\ \hline \end{array}$$

b.
$$\begin{array}{r} 6\ 7\ 8 \\ -2\ 5\ 7 \\ \hline \end{array}$$

c.
$$\begin{array}{r} 4\ 8\ 1 \\ +3\ 0\ 9 \\ \hline \end{array}$$

Lesson 42

Name _____

Part 1

a. _____ cents _____ cents

b. _____ cents _____ cents

Part 2

a. 73 – 10 = _____

b. 91 – 10 = _____

c. 436 – 10 = _____

d. 630 – 10 = _____

Part 3

a.
```
  1 8 5
– 1 6 1
```

b.
```
  1 9 9
– 1 9 6
```

c.
```
  7 8 6
– 7 0 4
```

Part 4

a. D > N > P

___ ___

b. P < D < Q

___ ___

Part 5

a.
```
  3
+ 6
```
b.
```
  5
+ 3
```
c.
```
  3
+ 4
```
d.
```
  7
– 3
```
e.
```
  6
– 3
```
f.
```
  8
+ 3
```
g.
```
  8
– 5
```
h.
```
  3
+ 3
```

i.
```
  9
– 3
```
j.
```
  3
+ 7
```
k.
```
 11
– 3
```
l.
```
 10
– 3
```
m.
```
  9
+ 3
```
n.
```
  4
– 3
```
o.
```
 12
– 9
```
p.
```
  3
+ 5
```

q.
```
 11
– 3
```
r.
```
  8
– 5
```
s.
```
  6
+ 3
```
t.
```
  7
– 3
```
u.
```
  5
+ 3
```
v.
```
 10
– 7
```
w.
```
  7
– 4
```
x.
```
  9
+ 3
```

Connecting Math Concepts

Lesson 42

Independent Work

Part 6

a. $35 + 10 =$ _____

b. $507 + 10 =$ _____

c. $66 + 10 =$ _____

d. $143 + 10 =$ _____

Part 7 | Write the 3 numbers from smallest to biggest.

a.
| 62 | 85 | 75 |

_____ _____ _____

b.
| 143 | 135 | 140 |

_____ _____ _____

c.
| 209 | 208 | 207 |

_____ _____ _____

Part 8

a. J is 18 more than P.
J is 59.
What number is P?   _____

⟶

b. T is 17 less than M.
T is 79.
What number is M?   _____

⟶

Part 9

a.
$$\begin{array}{r} 4\ 7\ 3 \\ -\ \ 7\ 1 \\ \hline \end{array}$$

b.
$$\begin{array}{r} 6\ 1\ 5 \\ +\ 1\ 4\ 5 \\ \hline \end{array}$$

c.
$$\begin{array}{r} 6\ 5\ 3 \\ -\ 2\ 4\ 3 \\ \hline \end{array}$$

Part 10 | Write the sign >, <, or =.

a. $20 + 5$ ▢ 26

b. 48 ▢ $49 - 1$

Lesson 43

Name _____

Part 1

a. $\begin{array}{r} 68 \\ -62 \\ \hline \end{array}$

b. $\begin{array}{r} 80 \\ -40 \\ \hline \end{array}$

c. $\begin{array}{r} 875 \\ -803 \\ \hline \end{array}$

d. $\begin{array}{r} 143 \\ -120 \\ \hline \end{array}$

e. $\begin{array}{r} 452 \\ -251 \\ \hline \end{array}$

Part 2 Write the letter **R**, **T**, or **C** in each shape.

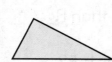

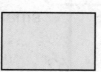

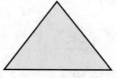

Part 3

a. 380 + 10 = _____

b. 380 + 100 = _____

c. 380 – 10 = _____

d. 380 – 100 = _____

e. 451 + 100 = _____

f. 451 + 10 = _____

g. 451 – 10 = _____

h. 451 – 100 = _____

Part 4

a. _____ cents   _____ cents

b. _____ cents   _____ cents

Connecting Math Concepts

Lesson 43

Name _____

Part 5

a.   11   7   9   b.   6   4   10

____   ____   ____   ____   ____   ____

     ____   ____   ____   ____

Part 6

a.	b.	c.	d.	e.	f.	g.	h.
3 + 8	8 − 3	9 + 3	8 + 3	6 − 3	3 + 6	3 + 3	9 − 3

i.	j.	k.	l.	m.	n.	o.	p.
3 + 5	10 − 3	7 − 4	3 + 4	11 − 3	3 + 7	10 − 7	5 + 3

q.	r.	s.	t.	u.	v.	w.	x.
3 + 7	6 − 3	9 − 6	3 + 8	3 + 4	6 + 3	8 + 3	8 − 3

y.	z.	A.	B.	C.	D.
11 − 3	3 + 3	7 − 4	3 + 6	4 + 3	8 − 5

Independent Work

Part 7 | Measure the line to each arrow.

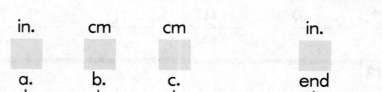

in.   cm   cm   in.

a.   b.   c.   end

Lesson 43

Name _____

Part 8 Write the missing numbers.

a. _27_ _36_ ____ ____ ____ ____ ____

b. _16_ _20_ ____ ____ ____ ____ ____

Part 9 Write 2 subtraction facts.

a. ■ —3→ 11

b. ■ —3→ 7

c. ■ —3→ 4

_____ _____ _____

_____ _____ _____

d. ■ —3→ 9

e. ■ —3→ 6

f. ■ —3→ 10

_____ _____ _____

_____ _____ _____

Part 10

a.
```
  2 6 0
-   4 0
```

b.
```
  6 1 8
+ 3 6 2
```

c.
```
  5 8 5
- 4 7 3
```

Part 11

_____ dollars _____ cents

Part 12

a.
```
  11
-  2
```

b.
```
   2
+  9
```

c.
```
   1
+  8
```

d.
```
  15
- 10
```

e.
```
   6
+  2
```

f.
```
   8
-  2
```

g.
```
   1
+  6
```

h.
```
   5
+  2
```

Part 13

a. _____ = 304

b. 400 + 90 + 0 = _____

Lesson 44

Name _____

Part 1 Write the letter **R**, **T**, or **C** in each shape.

1.

2.

3.

4.

5.

6.

7.

8.

Part 2

a. 9 ———3——→ _____

b. 9 ———8——→ _____

c. 9 ———5——→ _____

d. 9 ———7——→ _____

e. 9 ———9——→ _____

f. 9 ———1——→ _____

g. 9 ———4——→ _____

h. 9 ———2——→ _____

i. 9 ———6——→ _____

Part 3

a.

H > C
C > T

b.

P < J
J < R

Independent Work

Part 4 Measure the line to each arrow.

in.   in.   cm   cm

a.   b.   c.   end

Lesson 44

Name _____

Part 5 Write the cents for each side. Then make the sign >, <, or =.

a. _____ cents  _____ cents

  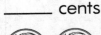

b. _____ cents _____ cents

c. _____ cents _____ cents

d. _____ cents _____ cents

Part 6

a. 10 + 367 = _____

b. 418 + 10 = _____

c. 103 + 10 = _____

d. 10 + 412 = _____

Part 7

a. 10 x 6 = _____

b. 4 x 3 = _____

c. 2 x 5 = _____

d. 5 x 4 = _____

Connecting Math Concepts

Lesson 44

Part 8

a.
$$\begin{array}{r} 5\ 7\ 3 \\ -\ 5\ 6\ 2 \\ \hline \end{array}$$

b.
$$\begin{array}{r} 4\ 6\ 2 \\ +\ 2\ 2\ 9 \\ \hline \end{array}$$

c.
$$\begin{array}{r} 6\ 2\ 7 \\ -\ 4\ 1\ 7 \\ \hline \end{array}$$

Part 9 Write 2 subtraction facts.

a. ■ —3→ 7

_____

_____

b. ■ —3→ 11

_____

_____

c. ■ —3→ 6

_____

_____

d. ■ —3→ 8

_____

_____

e. ■ —3→ 10

_____

_____

f. ■ —3→ 9

_____

_____

Part 10 Write the missing numbers.

a. __54__ __63__ ____ ____ ____

b. __16__ __20__ ____ ____ ____ ____ ____

Part 11

a.
$$\begin{array}{r} 9 \\ -\ 3 \\ \hline \end{array}$$

b.
$$\begin{array}{r} 4 \\ +\ 2 \\ \hline \end{array}$$

c.
$$\begin{array}{r} 3 \\ +\ 6 \\ \hline \end{array}$$

d.
$$\begin{array}{r} 5 \\ +\ 4 \\ \hline \end{array}$$

e.
$$\begin{array}{r} 19 \\ -\ 9 \\ \hline \end{array}$$

f.
$$\begin{array}{r} 10 \\ -\ 3 \\ \hline \end{array}$$

g.
$$\begin{array}{r} 6 \\ -\ 3 \\ \hline \end{array}$$

h.
$$\begin{array}{r} 9 \\ +\ 3 \\ \hline \end{array}$$

Lesson 45

Name _____

Part 1

a.  + =

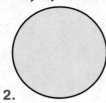

b.  + =

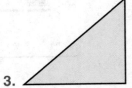

Part 2 Write the letter **R**, **T**, or **C** in each shape.

1.

2.

3.

4.

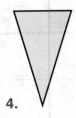

5.

6.

7.

8.

Part 3

a. $9 \xrightarrow{5}$ _____

_____

_____

b. $9 \xrightarrow{8}$ _____

_____

_____

c. $9 \xrightarrow{4}$ _____

_____

_____

d. $9 \xrightarrow{9}$ _____

_____

_____

e. $9 \xrightarrow{6}$ _____

_____

_____

Connecting Math Concepts

Lesson 45

Part 4

a. 8 > K
    K > 3

b. 7 < B
    B < 9

Part 5

a. 5 1
   − 3 3

b. 9 0
   − 2 8

c. 7 2
   − 1 9

Independent Work

Part 6

a. 7 6
   − 4 3

b. 4 2
   + 2 9

c. 5 2 6
   − 5 0 4

d. 2 8 2
   + 7 0 8

e. 8 9 3
   − 8 6 3

f. 7 5 3
   − 3 4 1

Part 7 Write 2 subtraction facts.

a. ■ ──3──▸ 12

_____

_____

b. ■ ──3──▸ 8

_____

_____

Lesson 45

Name _____

Part 8

a. V is 15 more than A.
V is 29.
What number is A?

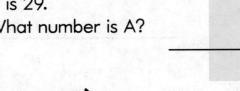

_____

⟶

b. C is 81 less than 291.
What number is C?

_____

⟶

Part 9

a. 12
− 9

b. 8
− 3

c. 7
+ 3

d. 3
+ 5

e. 2
+ 7

f. 11
− 8

g. 11
− 9

h. 8
− 6

i. 10
− 7

j. 3
+ 9

Part 10 Fill in the missing numbers.

a. __4__ __6__ ____ ____ ____

b. __24__ __28__ ____ ____ ____

c. __35__ __40__ ____ ____ ____

Connecting Math Concepts

Lesson

Name _____

Part 1

a.  =   b.   =

c. 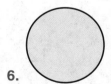 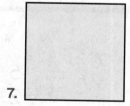 =

Part 2

1.   2.    3.    4.

5.   6.    7.   8.

Part 3

a. 9 + 7 = ____

b. 9 + 3 = ____

c. 9 + 9 = ____

d. 9 + 5 = ____

e. 9 + 2 = ____

f. 9 + 6 = ____

g. 9 + 1 = ____

h. 9 + 8 = ____

i. 9 + 4 = ____

Lesson 46

Name _____

Part 4

a.
```
  4 0
- 2 1
```

b.
```
  8 2
- 1 3
```

c.
```
  6 1
- 3 9
```

Independent Work

Part 5

a.

M > R
R > 3

_____   _____

_____

b.

G < F
F < 10

_____   _____

_____

Part 6

a.
```
  11
-  9
```

b.
```
   8
-  3
```

c.
```
   5
-  2
```

d.
```
  12
-  9
```

e.
```
  10
-  3
```

f.
```
   8
-  3
```

g.
```
   9
-  3
```

h.
```
  10
-  7
```

i.
```
   4
-  3
```

j.
```
   9
-  6
```

Part 7 | Measure the line to each arrow.

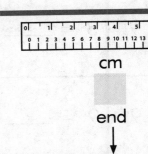

in.

in.   cm   cm

a.   b.   c.   end

Lesson 46

Name _____

Part 8 | Write the missing numbers.

a. __12__ __14__ ____ ____ ____ ____ ____

b. __30__ __40__ ____ ____ ____ ____ ____

c. __20__ __24__ ____ ____ ____ ____ ____

Part 9 | Make the sign >, <, or =.

a. 5 + 2 ⬜ 9

b. 20 + 10 ⬜ 40

c. _____ cents ⬜ _____ cents

d. _____ cents ⬜ _____ cents

Lesson 47

Name _____

a. 19 =

b. 35  =

Part 2 Write **R, T,** or **C** in each shape. Write **S** in each square.

1.

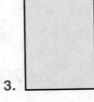
2.

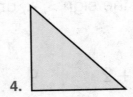
3.

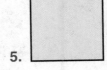
4.

5.

6.

7.

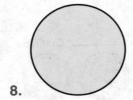
8.

9.

Part 3

a. 9 + 2 = _____

d. 9 + 7 = _____

g. 9 + 5 = _____

b. 9 + 8 = _____

e. 9 + 1 = _____

h. 9 + 9 = _____

c. 9 + 4 = _____

f. 9 + 3 = _____

i. 9 + 6 = _____

Connecting Math Concepts

Lesson 47

Part 4

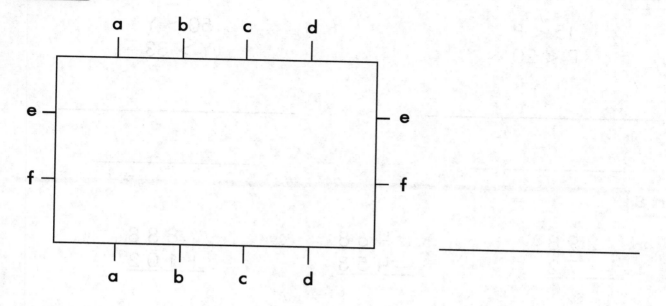

Part 5

a.
```
  5 1
- 3 2
```

b.
```
  9 0
- 1 1
```

c.
```
  3 2
- 2 9
```

Independent Work

Part 6 | Make the sign >, <, or =.

a. 2 x 7 [ ] 15

b. 20 [ ] 30 – 10

c. 63 [ ] 60 + 4

d. _____ cents [ ] _____ cents

Lesson 47

Name _____

Part 7

a. $15 < P$
    $P < 20$

b. $50 > Y$
    $Y > 33$

Part 8

a.
```
  298
-  72
```

b.
```
  468
- 453
```

c.
```
  888
+ 102
```

d.
```
  243
+ 339
```

e.
```
  518
- 205
```

f.
```
  236
+ 353
```

Part 9

a.
```
  12
-  9
```

b.
```
   5
-  3
```

c.
```
   7
-  4
```

d.
```
   8
-  6
```

e.
```
  10
-  7
```

f.
```
   8
-  5
```

g.
```
  12
-  3
```

h.
```
   5
-  2
```

i.
```
  10
-  3
```

j.
```
   9
-  6
```

Part 10 Complete each place-value fact.

a. _____ = 408

b. $300 + 10 + 6 =$ _____

c. _____ = 170

d. $200 + 90 + 0 =$ _____

Connecting Math Concepts

Lesson

Part 1

a. 27  =

  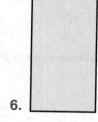

b. 18

c. 73   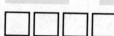 =

Part 2

1.

2.

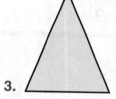

3.

4.

5.

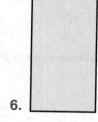

6.

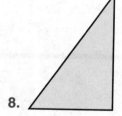

7.

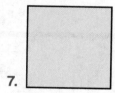

8.

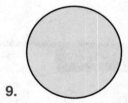

9.

Part 3

a. Sam is 7 years younger than Jan.

b. The plane is 3 years older than the truck.

c. Tina is 16 inches taller than Dan.

Lesson 48

Name _____

Part 4

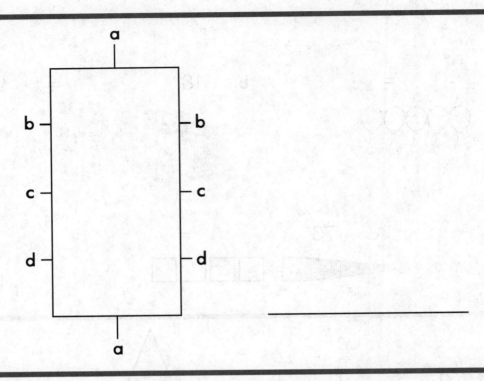

Part 5

a. $\begin{array}{r} 80 \\ -23 \\ \hline \end{array}$

b. $\begin{array}{r} 41 \\ -19 \\ \hline \end{array}$

c. $\begin{array}{r} 71 \\ -32 \\ \hline \end{array}$

Part 6

a. $\begin{array}{r} 3 \\ +9 \\ \hline \end{array}$
b. $\begin{array}{r} 9 \\ +4 \\ \hline \end{array}$
c. $\begin{array}{r} 9 \\ +3 \\ \hline \end{array}$
d. $\begin{array}{r} 14 \\ -5 \\ \hline \end{array}$
e. $\begin{array}{r} 13 \\ -3 \\ \hline \end{array}$
f. $\begin{array}{r} 5 \\ +9 \\ \hline \end{array}$
g. $\begin{array}{r} 12 \\ -3 \\ \hline \end{array}$

h. $\begin{array}{r} 14 \\ -4 \\ \hline \end{array}$
i. $\begin{array}{r} 13 \\ -4 \\ \hline \end{array}$
j. $\begin{array}{r} 9 \\ +4 \\ \hline \end{array}$
k. $\begin{array}{r} 4 \\ +9 \\ \hline \end{array}$
l. $\begin{array}{r} 14 \\ -4 \\ \hline \end{array}$
m. $\begin{array}{r} 9 \\ +2 \\ \hline \end{array}$
n. $\begin{array}{r} 5 \\ +9 \\ \hline \end{array}$

o. $\begin{array}{r} 9 \\ +4 \\ \hline \end{array}$
p. $\begin{array}{r} 12 \\ -2 \\ \hline \end{array}$
q. $\begin{array}{r} 2 \\ +9 \\ \hline \end{array}$
r. $\begin{array}{r} 14 \\ -5 \\ \hline \end{array}$
s. $\begin{array}{r} 12 \\ -3 \\ \hline \end{array}$
t. $\begin{array}{r} 9 \\ +5 \\ \hline \end{array}$

Connecting Math Concepts

Lesson 48

Independent Work

Part 7 Write the missing numbers.

a. __15__ __20__ ____ ____ ____ ____

b. __18__ __20__ __22__ ____ ____ ____

c. __16__ __20__ ____ ____ ____ ____

Part 8 Make the sign >, <, or =.

a. 20 5 × 5

b. 30 + 6 36

c. 9 20 − 10

d. ____ cents ____ cents

Part 9

a.

13 > R
R > 10

_____ _____

_____

b.

60 > P
P > 55

_____ _____

_____

Lesson 49

Name _____

Part 1

a. 28

b. 16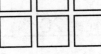

c. 39

d. 55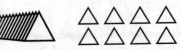

Part 2

a. Millie is 5 inches shorter than Paul.

b. Jake is 11 inches taller than Bill.

Part 3

a.
```
  7 2
- 4 3
```

b.
```
  7 2
- 4 1
```

c.
```
  8 9
- 1 7
```

d.
```
  8 0
- 1 7
```

e.
```
  6 1
- 2 9
```

Part 4

a. T > G

P > T

___ ___ ___

b. 11 < 14

9 < 11

___ ___ ___

Connecting Math Concepts

Lesson 49

Name _____

a. 6
  + 9

b. 9
  + 7

c. 17
  − 8

d. 11
  − 2

e. 9
  + 8

f. 17
  − 7

g. 15
  − 5

h. 15
  − 6

i. 9
  + 6

j. 8
  + 9

k. 7
  + 9

l. 16
  − 6

m. 11
  − 1

n. 16
  − 7

o. 8
  + 9

p. 15
  − 6

q. 9
  + 7

r. 17
  − 8

s. 15
  − 5

t. 6
  + 9

Independent Work

Part 6 Write **R**, **T**, or **C** in each shape. Then write **S** in each square.

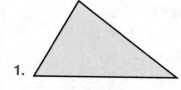

1.

2.

3.

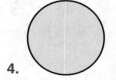

4.

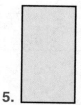

5.

6.

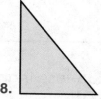

7.

8.

9.

Lesson 49

Name _____

Part 7

a. 13
  − 3

b.   7
  − 4

c.   7
  − 5

d. 10
  − 9

e. 11
  − 8

f. 10
  − 3

g. 12
  − 9

h.   7
  − 6

i.   4
  − 3

j. 11
  − 9

Part 8 | Measure the line to each arrow.

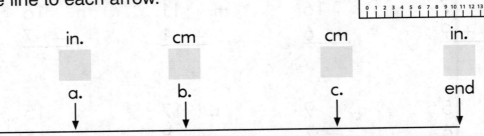

in.   cm   cm   in.

a.   b.   c.   end

Part 9 | Complete each place-value fact.

a. _____ = 618

b. 200 + 70 + 0 = _____

c. _____ = 111

d. 100 + 0 + 6 = _____

Part 10

a. 146 is 23 more than K.
   What number is K?

b. J is 51 more than P.
   P is 119.
   What number is J?

c. B is 235 less than T.
   T is 568.
   What number is B?

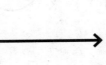

Lesson 50

Name _____

Part 1

a. <u>Millie</u> is 4 pounds heavier than <u>Jan</u>.

b. The <u>boat</u> is 200 pounds lighter than the <u>car</u>.

Part 2

a. $\begin{array}{r} 62 \\ -11 \\ \hline \end{array}$

b. $\begin{array}{r} 61 \\ -12 \\ \hline \end{array}$

c. $\begin{array}{r} 70 \\ -23 \\ \hline \end{array}$

d. $\begin{array}{r} 92 \\ -83 \\ \hline \end{array}$

e. $\begin{array}{r} 93 \\ -82 \\ \hline \end{array}$

Part 3

a. $14 < T$
$P < 14$

b. $20 > T$
$T > 10$

c. $R > M$
$P > R$

___ ___ ___   ___ ___ ___   ___ ___ ___

Part 4

a. $\begin{array}{r} 4 \\ +9 \\ \hline \end{array}$

b. $\begin{array}{r} 19 \\ -9 \\ \hline \end{array}$

c. $\begin{array}{r} 1 \\ +9 \\ \hline \end{array}$

d. $\begin{array}{r} 12 \\ -3 \\ \hline \end{array}$

e. $\begin{array}{r} 18 \\ -8 \\ \hline \end{array}$

f. $\begin{array}{r} 10 \\ -1 \\ \hline \end{array}$

g. $\begin{array}{r} 9 \\ +1 \\ \hline \end{array}$

h. $\begin{array}{r} 18 \\ -9 \\ \hline \end{array}$

i. $\begin{array}{r} 9 \\ +4 \\ \hline \end{array}$

j. $\begin{array}{r} 13 \\ -4 \\ \hline \end{array}$

k. $\begin{array}{r} 9 \\ +9 \\ \hline \end{array}$

l. $\begin{array}{r} 13 \\ -3 \\ \hline \end{array}$

m. $\begin{array}{r} 18 \\ -9 \\ \hline \end{array}$

n. $\begin{array}{r} 1 \\ +9 \\ \hline \end{array}$

o. $\begin{array}{r} 10 \\ -1 \\ \hline \end{array}$

p. $\begin{array}{r} 3 \\ +9 \\ \hline \end{array}$

q. $\begin{array}{r} 12 \\ -2 \\ \hline \end{array}$

r. $\begin{array}{r} 19 \\ -9 \\ \hline \end{array}$

s. $\begin{array}{r} 9 \\ +3 \\ \hline \end{array}$

t. $\begin{array}{r} 11 \\ -2 \\ \hline \end{array}$

Lesson 50

Name _____

Part 5

a. $\begin{array}{r} 2 \\ +\ 7 \\ \hline \end{array}$
b. $\begin{array}{r} 2 \\ +\ 9 \\ \hline \end{array}$
c. $\begin{array}{r} 3 \\ +\ 6 \\ \hline \end{array}$
d. $\begin{array}{r} 3 \\ +\ 9 \\ \hline \end{array}$
e. $\begin{array}{r} 9 \\ +\ 5 \\ \hline \end{array}$
f. $\begin{array}{r} 9 \\ +\ 3 \\ \hline \end{array}$
g. $\begin{array}{r} 12 \\ -\ 3 \\ \hline \end{array}$

h. $\begin{array}{r} 10 \\ -\ 3 \\ \hline \end{array}$
i. $\begin{array}{r} 4 \\ -\ 3 \\ \hline \end{array}$
j. $\begin{array}{r} 9 \\ -\ 6 \\ \hline \end{array}$
k. $\begin{array}{r} 10 \\ -\ 7 \\ \hline \end{array}$
l. $\begin{array}{r} 5 \\ -\ 5 \\ \hline \end{array}$
m. $\begin{array}{r} 12 \\ -\ 3 \\ \hline \end{array}$

Part 6 | Make the sign >, <, or =.

a. _____ cents _____ cents

b. $50 + 6$ 58 c. 83 $70 + 8$

Part 7

a. $\begin{array}{r} 5\ 8 \\ -5\ 2 \\ \hline \end{array}$
b. $\begin{array}{r} 1\ 3\ 2 \\ -\ \ \ 1\ 2 \\ \hline \end{array}$
c. $\begin{array}{r} 3\ 8\ 9 \\ -1\ 6\ 2 \\ \hline \end{array}$

Part 8 | Complete each place-value fact.

a. _____ = 308   b. _____ = 318

Part 9

a. $56 + 10 =$ _____   b. $149 + 10 =$ _____

c. $10 + 64 =$ _____   d. $10 + 143 =$ _____

Lesson

Part 1

a. M > V   _____ _____ _____
     15 > M

            _____

b. 25 < T   _____

      T < 27

                     _____

c. R < 13 _____
    10 < R

             _____

Part 2

a. <u>J</u>im was 14 pounds lighter than <u>H</u>eidi.

b. The <u>c</u>ar was 11 years older than the <u>t</u>ruck.

Part 3

a. 18   +   ☐   =   ☐

b. 28   +   ☐   =   ☐

Part 4

a. 7 + 9 = _____

b. 14 − 4 = _____

c. 12 − 2 = _____

d. 9 + 6 = _____

e. 16 − 6 = _____

f. 16 − 7 = _____

g. 14 − 5 = _____

h. 5 + 9 = _____

i. 15 − 5 = _____

j. 12 − 3 = _____

k. 5 + 9 = _____

l. 9 + 6 = _____

m. 15 − 6 = _____

n. 14 − 5 = _____

o. 6 + 9 = _____

p. 14 − 4 = _____

q. 15 − 5 = _____

r. 7 + 9 = _____

s. 9 + 5 = _____

t. 16 − 7 = _____

Lesson 51

Name _____

Independent Work

Part 5

a. 8 1
   − 3 2

b. 6 0
   − 3 3

c. 7 2
   − 3 1

Part 6 Measure the line to each arrow.

in.
   a.

cm
   b.

cm
   c.

in.
   end

Part 7

a. 8 2
   − 1 3

b. 5 7
   + 3 2

c. 6 9
   − 2 2

d. 9 1
   − 3 8

Part 8 Write the dollars and cents for each row.

a.

_____ dollars   _____ cents

b.

_____ dollars   _____ cents

Part 9

a. 9
   + 6

b. 5
   + 3

c. 2
   + 8

d. 9
   + 7

e. 4
   + 3

f. 9
   + 9

g. 12
   − 2

h. 14
   − 10

i. 11
   − 9

j. 10
   − 3

k. 13
   − 3

l. 7
   − 3

m. 12
   − 3

Connecting Math Concepts

Lesson

Part 1

a. V > 10 _____

    18 > V

    _____

b. P < T _____

    T < 5

    _____

c. 10 > P _____

    Y > 10

    _____

Part 2

a. 39 + ▢ = ▢

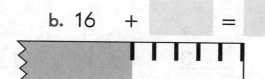

b. 16 + ▢ = ▢

Part 3

a. 9 + 6	**b.** 9 + 9	**c.** 17 − 7	**d.** 9 + 3	**e.** 9 + 8
f. 4 + 9	**g.** 13 − 4	**h.** 17 − 8	**i.** 14 − 4	**j.** 9 + 7
k. 8 + 9	**l.** 17 − 7	**m.** 13 − 4	**n.** 9 + 9	**o.** 18 − 9
p. 9 + 6	**q.** 14 − 4	**r.** 13 − 4	**s.** 14 − 5	**t.** 9 + 3

Lesson 52

Independent Work

Part 4 Figure out the missing numbers.

a. 56 + ▢ = ▢

b. 18 + ▢ = ▢

Part 5

a. 3
   + 5

b. 7
   + 3

c. 4
   + 2

d. 3
   + 7

e. 9
   + 3

f. 3
   + 8

g. 3
   + 2

h. 6
   + 9

i. 8
   + 9

j. 4
   + 9

k. 8
   + 10

l. 5
   + 9

m. 6
   + 3

n. 7
   + 9

Part 6

a. 40
   − 31

b. 88
   − 32

c. 72
   − 23

Lesson 52

Name _____

Part 7 Write the dollars and cents for each row.

a.

_____ dollars   _____ cents

---

b.

_____ dollars   _____ cents

---

c.

_____ dollars   _____ cents

Part 8

a. $132 + 10 =$ _____

b. $10 + 207 =$ _____

c. $10 + 86 =$ _____

Part 9

a.
$$\begin{array}{r} 61 \\ -59 \\ \hline \end{array}$$

b.
$$\begin{array}{r} 124 \\ +573 \\ \hline \end{array}$$

c.
$$\begin{array}{r} 581 \\ -463 \\ \hline \end{array}$$

Lesson

Name _____

Part 1

a. 20 + 10 = _____   e. 80 + 10 = _____

b. 10 + 26 = _____   f. 10 + 87 = _____

c. 10 + 75 = _____   g. 32 + 10 = _____

d. 18 + 10 = _____

Part 2

a. M < Y
   Y < 99

   _____

b. N > F
   F > P

   _____

Part 3

a. 41 + [____] = [____]

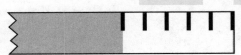

b. 26 + [____] = [____]

Part 4

a.	b.	c.	d.	e.	f.	g.	h.
16	6	9	15	14	9	13	16
− 6	+ 9	+ 8	− 6	− 5	+ 5	− 3	− 7

i.	j.	k.	l.	m.	n.	o.	p.
3	9	9	11	7	13	15	9
+ 9	+ 2	+ 3	− 1	+ 9	− 4	− 5	+ 4

q.	r.	s.	t.	u.	v.	w.	x.
5	14	11	17	9	12	8	9
+ 9	− 4	− 2	− 7	+ 7	− 2	+ 9	+ 6

y.	z.	A.	B.	C.	D.
4	17	18	12	18	9
+ 9	− 8	− 9	− 3	− 10	+ 9

Connecting Math Concepts

Lesson 53

Name _____

Independent Work

Part 5

a.   9
  + 5

b.   4
  + 9

c.   7
  + 9

d.   3
  + 9

e. 13
  − 10

f. 12
  − 3

g. 10
  − 2

h. 11
  − 10

i. 11
  − 2

j. 10
  − 3

k. 11
  − 3

l.   9
  − 6

Part 6 | Measure the line to each arrow.

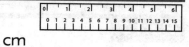

cm

in.

in.

cm

a.

b.

c.

end

Part 7 | Make the sign >, <, or =.

a. _____ cents _____ cents

b. 23 [ ] 10 + 12

Part 8

a.   3 2 5
  + 5 6 3

b.   8 4 0
  − 8 0 7

c.   6 5 2
  − 6 4 9

Lesson

Name _____

Part 1

a. 59 + 10 = ____   e. 10 + 89 = ____

b. 10 + 40 = ____   f. 60 + 10 = ____

c. 47 + 10 = ____   g. 13 + 10 = ____

d. 10 + 65 = ____

Part 2

a. 11 < P   b. T < 15   c. 9 > 6   d. 5 > J
   P < 15   F < T   W > 9   J > E

_____   _____   _____   _____

Part 3

a. 37   +   [   ]   =   [   ]   b. 65   +   [   ]   =   [   ]

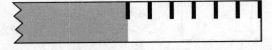   

Part 4

1.   2.   3.   4.

5.   6.   7.   8.

9.   10.   11.   12.

- triangle _____   • sphere _____   • square _____

- pyramid _____   • circle _____   • cube _____

Connecting Math Concepts

Lesson 54

Name _____

a. 11 − 2	b. 15 − 5	c. 9 + 5	d. 18 − 9	e. 6 + 9	f. 13 − 3	g. 16 − 6	h. 9 + 4
i. 13 − 4	j. 9 + 8	k. 9 + 7	l. 11 − 1	m. 7 + 9	n. 14 − 5	o. 4 + 9	p. 12 − 3
q. 9 + 3	r. 16 − 7	s. 14 − 4	t. 3 + 9	u. 8 + 9	v. 9 + 2	w. 12 − 2	x. 15 − 6
y. 9 + 6	z. 17 − 7	A. 5 + 9	B. 9 + 9	C. 17 − 8	D. 18 − 8		

Independent Work

Part 6

a. 5 6 8
  − 3 3 6

b. 5 4 0
  + 2 3 7

c. 7 5 8
  − 1 2 5

Part 7 Complete each place-value fact.

a. _____ = 711   b. _____ = 502

Lesson 54

Name _____

Part 8 Measure the line to each arrow.

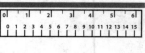

cm	in.	in.	cm

a. b. c. end

↓   ↓   ↓   ↓   ↓

Part 9

a. 4 6
   – 2 5

b. 6 1
   – 4 3

c. 8 2
   – 6 9

Part 10 Write the dollars and cents for each row.

a.

_____ dollars   _____ cents

b.

_____ dollars   _____ cents

Lesson

Part 1

a. 5 + 5 = _____   d. 10 − 5 = _____   g. 4 + 4 = _____

b. 4 + 4 = _____   e. 8 − 4 = _____   h. 6 + 6 = _____

c. 6 + 6 = _____   f. 12 − 6 = _____   i. 5 + 5 = _____

Part 2

a. 12 + 10 = _____   e. 84 + 10 = _____

b. 66 + 10 = _____   f. 17 + 10 = _____

c. 25 + 10 = _____   g. 39 + 10 = _____

d. 50 + 10 = _____

Independent Work

Part 3 Write the statement without the middle value.

a. N > C C > V	b. P < T K < P	c. 13 < T R < 13
_____	_____	_____

Part 4

| a. 7 + 9 | b. 9 + 3 | c. 12 − 3 | d. 9 − 3 | e. 9 − 7 | f. 9 − 6 | g. 9 + 5 |

Lesson 55

Name _____

Part 5

a. 10 + 481 = _____

b. 427 + 10 = _____

c. 10 + 86 = _____

d. 89 + 10 = _____

Part 6 | Write the sign >, <, or =.

a. _____ cents 32 cents   b. _____ cents 60 cents

Part 7

a. 4 6 8
   + 1 9

b. 2 5 3
   − 1 3 4

c. 8 8 0
   − 8 3 9

d. 5 6 3
   + 4 1 9

Part 8 | Figure out the missing numbers.

a. 33 + ▢ = ▢

Connecting Math Concepts

Lesson

Name _____

Part 1

a. $14 - 7 =$ _____

b. $4 + 4 =$ _____

c. $6 + 6 =$ _____

d. $16 - 8 =$ _____

e. $5 + 5 =$ _____

f. $12 - 6 =$ _____

g. $8 + 8 =$ _____

h. $7 + 7 =$ _____

i. $8 - 4 =$ _____

j. $6 + 6 =$ _____

k. $5 + 5 =$ _____

l. $8 + 8 =$ _____

Part 2

1.

2.

3.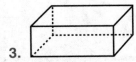

4.

5.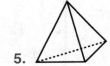

6.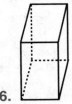

• cubes _____

• pyramids _____

• rectangular prisms _____

Independent Work

Part 3

a. $55 + 10 =$ _____

b. $10 + 40 =$ _____

c. $10 + 13 =$ _____

d. $39 + 10 =$ _____

e. $27 + 10 =$ _____

f. $10 + 19 =$ _____

g. $72 + 10 =$ _____

Lesson 56

Name _____

Part 4

a.	b.	c.	d.	e.	f.	g.
7 + 2	11 − 9	13 − 10	12 − 2	9 − 6	9 + 6	5 + 2

Part 5

a. J < P
    P < Z

    _____

b. N < T
    10 < N

    _____

c. B < 18
    18 < K

    _____

Part 6

a.	b.	c.	d.
678 − 609	450 − 231	263 + 526	408 + 253

Part 7   Write the sign >, <, or =.

    a. 36 + 10 ☐ 46

    b. 59 ☐ 58 + 2

Part 8

a. 26 + ☐ = ☐

b. 33 + ☐ = ☐

Connecting Math Concepts

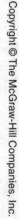

Lesson

Name _____

a. 3 + 3 = _____   k. 6 + 6 = _____

b. 8 + 8 = _____   l. 5 + 5 = _____

c. 10 + 10 = _____   m. 1 + 1 = _____

d. 7 + 7 = _____   n. 7 + 7 = _____

e. 5 + 5 = _____   o. 3 + 3 = _____

f. 9 + 9 = _____   p. 10 + 10 = _____

g. 2 + 2 = _____   q. 9 + 9 = _____

h. 6 + 6 = _____   r. 4 + 4 = _____

i. 1 + 1 = _____   s. 8 + 8 = _____

j. 4 + 4 = _____   t. 2 + 2 = _____

Part 2

 1.    2.   3.   4.

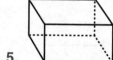 5.   6.   7.   8.

• rectangular prism _____

• square _____

• cube _____

• rectangle _____

Lesson 57

Name _____

Independent Work

Part 3

a. 12
  − 9

b. 14
  − 5

c.   5
  + 9

d.   3
  + 9

e.   7
  + 9

f. 12
  − 10

g.   6
  + 9

h.   9
  + 9

i. 10
  − 7

j. 13
  − 9

k.   9
  + 8

l. 10
  + 7

Part 4 Write the sign >, <, or =.

a. _____ cents   22 cents

b. _____ cents   61 cents

Part 5

a. _____ = 241

b. 400 + 20 + 9 = _____

c. _____ = 507

d. 700 + 90 + 0 = _____

Part 6

a. 678
  + 312

b. 840
  − 238

c. 357
  + 633

d. 856
  − 814

Lesson

Name _____

Part 1

a. 4 + 4 = ____	k. 8 + 8 = ____	u. 10 + 10 = ____
b. 10 + 10 = ____	l. 6 + 6 = ____	v. 14 − 7 = ____
c. 12 − 6 = ____	m. 2 + 2 = ____	w. 5 + 5 = ____
d. 2 + 2 = ____	n. 16 − 8 = ____	x. 6 + 6 = ____
e. 7 + 7 = ____	o. 5 + 5 = ____	y. 7 + 7 = ____
f. 8 + 8 = ____	p. 3 + 3 = ____	z. 8 − 4 = ____
g. 1 + 1 = ____	q. 4 + 4 = ____	A. 3 + 3 = ____
h. 14 − 7 = ____	r. 8 + 8 = ____	B. 1 + 1 = ____
i. 6 + 6 = ____	s. 12 − 6 = ____	C. 9 + 9 = ____
j. 9 + 9 = ____	t. 7 + 7 = ____	D. 4 + 4 = ____

Part 2

1.   2.   3.   4.

5.   6.   7.   8.

- cube _____
- rectangular prism _____
- rectangle _____
- square _____

Lesson 58

Name _____

Independent Work

Part 3

a. 4
+ 9

b. 15
− 6

c. 11
− 3

d. 2
+ 9

e. 17
− 8

f. 18
− 9

g. 9
+ 7

h. 9
+ 4

i. 6
− 4

j. 6
+ 4

k. 8
+ 9

l. 9
+ 5

Part 4   Write the sign >, <, or =.

a. _____ cents ▢ 38 cents

c. 14 + 10 ▢ 25

b. 85 cents ▢ _____ cents

d. 82 ▢ 10 + 71

Part 5

a. 526 + 10 = _____

b. 10 + 780 = _____

c. 137 + 10 = _____

d. 10 + 309 = _____

Part 6

a. 2 6 9
+ 5 2 4

b. 6 7 4
− 3 3 3

c. 4 3 2
+ 2 8

d. 5 8 3
− 1 3 9

Part 7

a. M > P
P > 43

_____

b. K > R
T > K

_____

c. 20 < J
V < 20

_____

Connecting Math Concepts

Lesson

Name _____

a. $\begin{array}{r} 10 \\ + 10 \\ \hline \end{array}$ b. $\begin{array}{r} 14 \\ - 7 \\ \hline \end{array}$ c. $\begin{array}{r} 3 \\ + 3 \\ \hline \end{array}$ d. $\begin{array}{r} 5 \\ + 5 \\ \hline \end{array}$ e. $\begin{array}{r} 4 \\ + 4 \\ \hline \end{array}$

f. $\begin{array}{r} 2 \\ + 2 \\ \hline \end{array}$ g. $\begin{array}{r} 6 \\ + 6 \\ \hline \end{array}$ h. $\begin{array}{r} 16 \\ - 8 \\ \hline \end{array}$ i. $\begin{array}{r} 7 \\ + 7 \\ \hline \end{array}$ j. $\begin{array}{r} 9 \\ + 9 \\ \hline \end{array}$

k. $\begin{array}{r} 5 \\ + 5 \\ \hline \end{array}$ l. $\begin{array}{r} 1 \\ + 1 \\ \hline \end{array}$ m. $\begin{array}{r} 8 \\ + 8 \\ \hline \end{array}$ n. $\begin{array}{r} 14 \\ - 7 \\ \hline \end{array}$ o. $\begin{array}{r} 6 \\ + 6 \\ \hline \end{array}$

p. $\begin{array}{r} 7 \\ + 7 \\ \hline \end{array}$ q. $\begin{array}{r} 2 \\ + 2 \\ \hline \end{array}$ r. $\begin{array}{r} 10 \\ + 10 \\ \hline \end{array}$ s. $\begin{array}{r} 16 \\ - 8 \\ \hline \end{array}$ t. $\begin{array}{r} 4 \\ + 4 \\ \hline \end{array}$

u. $\begin{array}{r} 8 \\ + 8 \\ \hline \end{array}$ v. $\begin{array}{r} 1 \\ + 1 \\ \hline \end{array}$ w. $\begin{array}{r} 12 \\ - 6 \\ \hline \end{array}$ x. $\begin{array}{r} 5 \\ + 5 \\ \hline \end{array}$ y. $\begin{array}{r} 9 \\ + 9 \\ \hline \end{array}$

z. $\begin{array}{r} 3 \\ + 3 \\ \hline \end{array}$ A. $\begin{array}{r} 4 \\ + 4 \\ \hline \end{array}$ B. $\begin{array}{r} 6 \\ + 6 \\ \hline \end{array}$ C. $\begin{array}{r} 8 \\ - 4 \\ \hline \end{array}$ D. $\begin{array}{r} 7 \\ + 7 \\ \hline \end{array}$

Lesson 59

Part 2

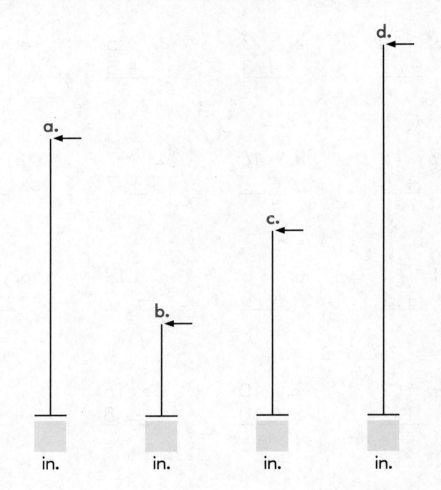

a. in.

b. in.

c. in.

d. in.

Independent Work

Part 3

a.	b.	c.	d.	e.	f.	g.	h.
2	5	9	9	15	2	9	9
+ 9	− 3	+ 8	+ 5	− 6	+ 9	− 6	+10

Part 4

a. 10 + 403 = _____

b. 421 + 10 = _____

c. 10 + 717 = _____

d. 706 + 10 = _____

Lesson 59

Name _____

Part 5 Complete each place-value fact.

a. _____ = 530   b. $400 + 0 + 6 =$ _____

c. _____ = 709   d. $900 + 10 + 1 =$ _____

e. _____ = 600   f. $900 + 0 + 0 =$ _____

Part 6

a.
$$\begin{array}{r} 2\,5\,7 \\ +\,7\,3\,2 \\ \hline \end{array}$$

b.
$$\begin{array}{r} 4\,8\,0 \\ -\,3\,2\,2 \\ \hline \end{array}$$

c.
$$\begin{array}{r} 6\,5\,1 \\ -\,6\,2\,9 \\ \hline \end{array}$$

d.
$$\begin{array}{r} 1\,6\,2 \\ +\,1\,2\,8 \\ \hline \end{array}$$

Part 7 Write the sign >, <, or =.

a. $19 - 10$ ☐ 10 b. 49 ☐ $9 + 40$

c. $8 + 10$ ☐ 18 d. 11 ☐ $19 - 9$

Part 8

a. 54 + ☐ = ☐

b. 86 + ☐ = ☐

Lesson 60

Part 1

a. 7 + 7 = _____
b. 5 + 5 = _____
c. 8 + 8 = _____
d. 14 − 7 = _____
e. 1 + 1 = _____
f. 10 + 10 = _____
g. 3 + 3 = _____
h. 12 − 6 = _____
i. 7 + 7 = _____
j. 8 + 8 = _____

k. 10 + 10 = _____
l. 7 + 7 = _____
m. 1 + 1 = _____
n. 9 + 9 = _____
o. 20 − 10 = _____
p. 6 + 6 = _____
q. 3 + 3 = _____
r. 4 + 4 = _____
s. 16 − 8 = _____
t. 5 + 5 = _____

u. 2 + 2 = _____
v. 9 + 9 = _____
w. 18 − 9 = _____
x. 2 + 2 = _____
y. 8 + 8 = _____
z. 6 + 6 = _____
A. 5 + 5 = _____
B. 10 − 5 = _____
C. 4 + 4 = _____
D. 6 + 6 = _____

Part 2

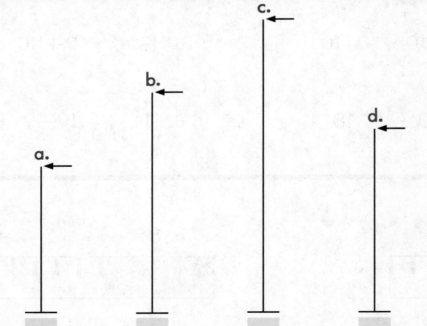

a. ____ cm
b. ____ cm
c. ____ cm
d. ____ cm

Connecting Math Concepts

Lesson 60

Name _____

Independent Work

Part 3

a.   48
   + 2 2

b.   79
   + 1 3

c.   87
   − 3 8

d.   59
   − 2 7

Part 4 Write the sign >, <, or =.

a. 68 dollars _____ _____ dollars

Part 5

a.   3
   + 9

b.   14
   − 5

c.   15
   − 10

d.   5
   + 9

e.   12
   − 9

f.   9
   − 3

Copyright © The McGraw-Hill Companies, Inc.

Connecting Math Concepts

Lesson 60 **157**

Lesson 61

Name _____

Part 1

a. 12
  − 9

b.   7
  + 7

c. 13
  − 10

d.   3
  + 9

e. 12
  − 3

f.   6
  − 6

g.   7
  − 4

h. 16
  − 6

i. 17
  − 8

j. 14
  − 5

k. 19
  − 10

l. 13
  − 4

m.   5
  + 9

n. 12
  − 6

o.   8
  + 9

p.   9
  − 3

Part 2

a. B ———→ F → R

   B = 10
   F = 20

b. P ———→ Q → T

   T = 12
   Q = 2

Part 3

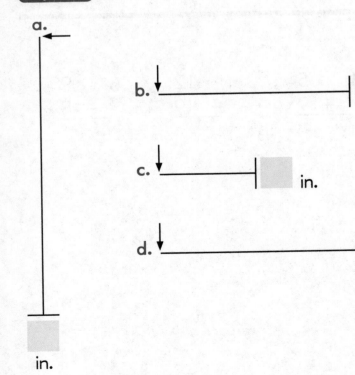

b. in.

c. in.

d. cm

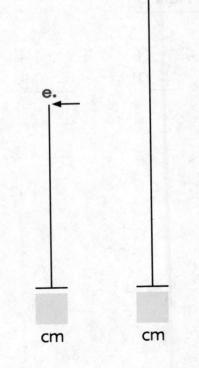

e. cm

f. cm

a. in.

Connecting Math Concepts

Lesson

Name _____

Part 4

a. 9 + ■ = 11

b. 4 + ■ = 7

c. 10 + ■ = 15

→   →   →

Independent Work

Part 5

a.   3 6
    + 4 6

b.   7 2
    − 3 6

c.   1 5
    + 6 5

d.   3 4
    − 2 7

Part 6 Write the sign >, <, or =.

a. _____ cents  70 cents

Part 7

a. C > J
    B > C

_____

b. 12 < K
    K < B

_____

c. T < 5
    M < T

_____

Part 8

a.   6
    + 6

b.   5
    + 5

c.   14
    − 7

d.   9
    − 3

e.   12
    − 3

f.   7
    + 3

g.   2
    + 9

h.   10
    − 2

Lesson 62

Name _____

Part 1

a. 14
   − 7

b. 18
   − 10

c. 17
   − 8

d. 15
   − 6

e.   7
   + 9

f.   4
   + 9

g. 14
   − 5

h. 10
   − 3

i.   8
   + 9

j.   6
   + 9

k.   8
   − 3

l. 12
   − 3

m.   7
   − 4

n.   8
   − 5

o. 11
   − 3

Part 2

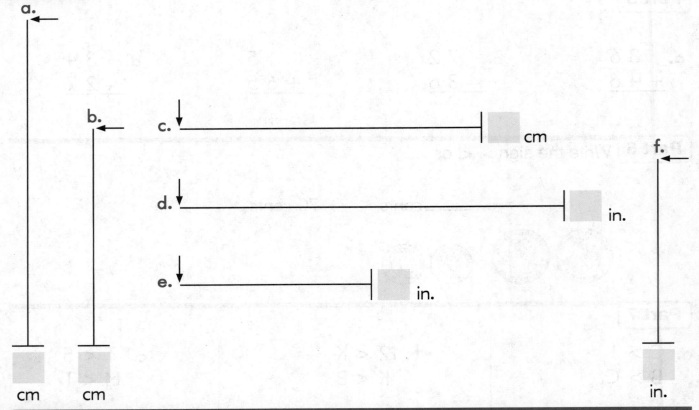

Part 3

a. 3 + ■ = 5

b. 4 + ■ = 8

c. 8 + ■ = 11

d. 2 + ■ = 12

e. 5 + ■ = 10

Connecting Math Concepts

Lesson 62

Name _____

Independent Work

Part 4

a.
```
   3 2 8
 + 2 5 8
```

b.
```
   8 2
 - 6 3
```

c.
```
   7 6
 - 3 8
```

d.
```
   6 5 4
 + 2 2 4
```

Part 5 | Write the sign >, <, or =.

```
   4 5
 + 4 5
```
a. ☐ 88

Part 6

a. M > 10
   10 > T

_____

b. B < 8
   5 < B

_____

c. P < K
   K < 15

_____

Part 7

a.
```
   5
 + 5
```
b.
```
   3
 + 9
```
c.
```
   3
 + 6
```
d.
```
   5
 + 3
```
e.
```
   9
 + 6
```
f.
```
   5
 + 9
```
g.
```
   7
 + 9
```
h.
```
   6
 + 6
```

i.
```
  10
 - 5
```
j.
```
   8
 - 4
```
k.
```
  14
 - 7
```
l.
```
   7
 - 3
```
m.
```
  12
 - 2
```
n.
```
  12
 - 6
```
o.
```
  11
 - 3
```
p.
```
  16
 - 8
```

Lesson 63

Name _____

Part 1

a. $10 - 3 =$ ___ e. $5 + 9 =$ ___ i. $15 - 6 =$ ___ m. $7 - 5 =$ ___

b. $7 - 4 =$ ___ f. $9 - 6 =$ ___ j. $13 - 3 =$ ___ n. $18 - 10 =$ ___

c. $9 + 9 =$ ___ g. $14 - 7 =$ ___ k. $7 + 9 =$ ___ o. $18 - 8 =$ ___

d. $16 - 8 =$ ___ h. $8 + 9 =$ ___ l. $3 + 9 =$ ___ p. $5 - 5 =$ ___

Part 2

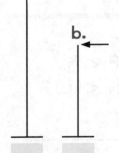

a.

b.

c. in.

d. cm

e. cm

f. cm

in.   in.

Part 3

a. $10 + \blacksquare = 14$ b. $9 + \blacksquare = 12$ c. $3 + \blacksquare = 5$

d. $2 + \blacksquare = 11$ e. $1 + \blacksquare = 8$

Connecting Math Concepts

Lesson

Part 4

a. 45 + 10 = ____   b. 74 + 10 = ____   c. 51 + 10 = ____

45 + 9 = ____   74 + 9 = ____   51 + 9 = ____

d. 16 + 10 = ____   e. 83 + 10 = ____   f. 18 + 10 = ____

16 + 9 = ____   83 + 9 = ____   18 + 9 = ____

Independent Work

Part 5

a. 392   b. 339   c. 235   d. 74
   −146   −136   +125   −17

Part 6 Write the sign >, <, or =.

                106
              + 16

a. 1 2 3

Part 7

a. 4   b. 12   c. 14   d. 4   e. 7   f. 5   g. 9   h. 9
   +9   − 3   − 5   +4   +7   + 5   + 6   + 8

Lesson 64

Name _____

Part 1

a. $8 - 3 =$ _____

b. $11 - 8 =$ _____

c. $14 - 7 =$ _____

d. $10 - 7 =$ _____

e. $6 + 9 =$ _____

f. $12 - 10 =$ _____

g. $16 - 8 =$ _____

h. $4 + 9 =$ _____

i. $3 + 9 =$ _____

j. $7 + 9 =$ _____

k. $12 - 6 =$ _____

l. $13 - 4 =$ _____

m. $2 + 9 =$ _____

n. $17 - 10 =$ _____

o. $14 - 7 =$ _____

p. $8 + 9 =$ _____

Part 2

cm

cm

a

b ① d

c

cm

cm

cm

b

a ② c

cm

d

cm

Part 3

a. $6 +$ ■ $= 12$

b. $2 +$ ■ $= 10$

c. $7 +$ ■ $= 17$

d. $4 +$ ■ $= 6$

e. $8 +$ ■ $= 11$

Connecting Math Concepts

Lesson 64

Name _____

Part 4

a. 31 + 10 = _____   b. 16 + 10 = _____   c. 88 + 10 = _____

    31 + 9 = _____   16 + 9 = _____   88 + 9 = _____

d. 12 + 10 = _____   e. 49 + 10 = _____   f. 67 + 10 = _____

    12 + 9 = _____   49 + 9 = _____   67 + 9 = _____

Independent Work

Part 5

a.   74
  + 19

b.   76
  − 58

c.   82
  − 36

d.   37
  + 47

Part 6 Write the sign >, <, or =.

a. $ _____ $50.00

Part 7

a. 30 > K
  T > 30

_____

b. P > 10
  10 > T

_____

c. 7 < T
  T < 12

_____

Part 8

a.   4
 + 4

b.   9
 − 2

c.   7
 − 3

d.   12
 − 3

e.   6
 + 6

f.   7
 + 7

g.   12
 − 6

h.   5
 + 5

Lesson 65

Name _____

Part 1

1. grapes cereal	2. ruler shovel	3. gorilla mouse	4. cheese tomato
5. airplane car	6. tractor bus	7. squirrel pigeon	8. wheelbarrow hammer

⟶ v   ⟶ f   ⟶ a   ⟶ t

⟶ v   ⟶ f   ⟶ a   ⟶ t

Part 2

a. $56 + 10 =$ _____

   $56 + 9 =$ _____

b. $42 + 10 =$ _____

   $42 + 9 =$ _____

c. $75 + 10 =$ _____

   $75 + 9 =$ _____

d. $31 + 10 =$ _____

   $31 + 9 =$ _____

e. $89 + 10 =$ _____

   $89 + 9 =$ _____

f. $66 + 10 =$ _____

   $66 + 9 =$ _____

Part 3

a. $3 +$ ▢ $= 10$ b. $8 +$ ▢ $= 16$ c. $10 +$ ▢ $= 20$

d. $4 +$ ▢ $= 7$ e. $2 +$ ▢ $= 12$

Lesson 65

Part 4

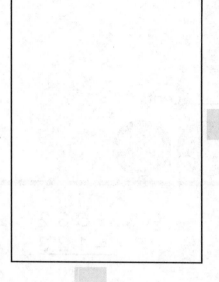

a.

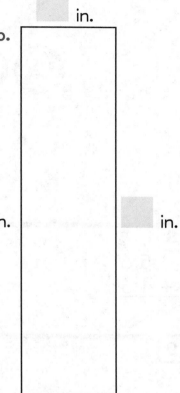

b.

Independent Work

Part 5

a.	b.	c.	d.	e.	f.	g.	h.
10	2	10	12	7	9	8	9
− 3	+ 9	− 5	− 9	+ 9	+ 9	+ 9	+ 10

Part 6

a. D > 5
   S > D

_____

b. K < F
   F < T

_____

c. R < 21
   15 < R

_____

Lesson 65

Part 7 Write the sign >, <, or =.

a. _____ cents 14 cents

b.   45 cents _____ cents

Part 8

a. 5 6
   + 1 6

b. 4 1 9
   + 6 9

c. 5 3 2
   − 1 2 3

Part 9

d. ←

a. ←

b. ↓ _____ ▮ in.

c. ↓ _____ ▮ cm

cm

in.

Lesson 66

Name _____

Part 1

a. 46 + 10 = ____

46 + 9 = ____

b. 58 + 10 = ____

58 + 9 = ____

c. 32 + 10 = ____

32 + 9 = ____

d. 88 + 10 = ____

88 + 9 = ____

e. 61 + 10 = ____

61 + 9 = ____

f. 25 + 10 = ____

25 + 9 = ____

Part 2

a.

8 cm

10 cm

____ cm

b.

____ cm

6 cm

6 cm

____ cm

____ cm

Part 3

a. 7 + ____ = 9

b. 6 + ____ = 12

c. 9 + ____ = 15

d. 7 + ____ = 17

e. 4 + ____ = 7

Lesson 66

Name _____

Part 4

| children | fruit | bugs | games | buildings |

| baseball tag | | pear orange | | school garage |

a. ⟶   b. ⟶   c. ⟶

| boy girl | | grasshopper ant |

d. ⟶   e. ⟶

Independent Work

Part 5

a. 12
− 6

b. 9
− 3

c. 5
+ 3

d. 10
− 3

e. 10
− 5

f. 3
+ 8

g. 5
+ 9

h. 3
+ 9

Part 6 Write the sign >, <, or =.

a. 62 cents ☐ _____ cents

b. _____ cents ☐ 16 cents

c. 85 cents ☐ _____ cents

Part 7

a. 248
+ 433

b. 135
+ 105

c. 546
− 527

Connecting Math Concepts

Lesson 67

Name _____

a.  + 3 = 50   b. 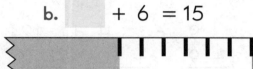 + 6 = 15

Part 2

a. 52 + 9 = _____   b. 35 + 9 = _____   c. 67 + 9 = _____

d. 11 + 9 = _____   e. 48 + 9 = _____   f. 39 + 9 = _____

Part 3

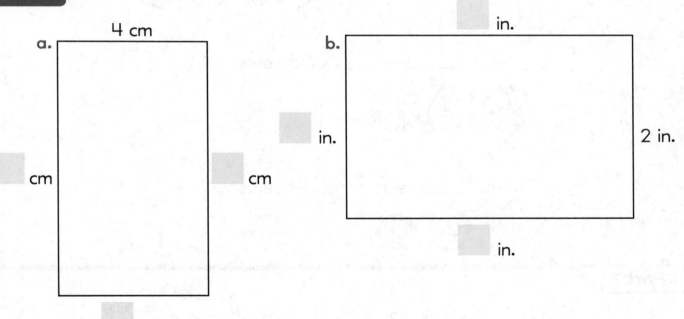

Part 4

a. 10 + ▢ = 18   b. 7 + ▢ = 14   c. 3 + ▢ = 10

d. 2 + ▢ = 11   e. 40 + ▢ = 45

Copyright © The McGraw-Hill Companies, Inc.

Lesson 67

Name _____

Part 5

a.	b.	c.	d.	e.	f.	g.	h.
4	9	4	12	16	20	14	5
+ 3	+ 9	+ 10	− 2	− 8	− 10	− 7	+ 5

Part 6 | Write the sign >, <, or =.

a.   47 cents ▢ ____ cents

b. ____ cents ▢ 75 cents

c. ____ cents ▢ 14 cents

Part 7

a. 8 6 2
 − 4 5 6

b. 3 9 0
 − 8 5

Part 8

a. K < 8
    8 < B

_____

b. M < 15
    15 < P

_____

c. T > P
    H > T

_____

Lesson 68

Name _____

Part 1

a. ▢ + 5 = 40

b. ▢ + 4 = 27

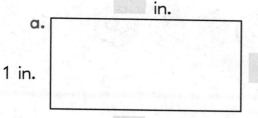

Part 2

a. 36 + 9 = ____

b. 71 + 9 = ____

c. 87 + 9 = ____

d. 42 + 9 = ____

e. 55 + 9 = ____

f. 63 + 9 = ____

Part 3

a.

▢ in.

1 in.

▢ in.

▢ in.

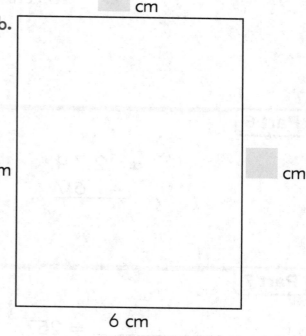

b.

▢ cm

▢ cm

▢ cm

6 cm

Independent Work

Part 4

a.	b.	c.	d.	e.	f.	g.	h.
17	14	18	4	8	6	5	9
− 9	− 7	− 9	+ 9	+ 8	+ 10	+ 9	+ 3

Lesson 68

Name _____

Part 5 | Write the sign >, <, or =.

a. 62 cents [ ] _____ cents

b. _____ cents [ ] 36 cents

c. 46 cents [ ] _____ cents

Part 6

a.
```
  2 7 4
-   6 7
```

b.
```
  6 8 6
- 3 5 8
```

Part 7

a. _____ = 257

b. 42 = _____

c. 190 = _____

d. _____ = 88

e. _____ = 700

f. 900 = _____

Connecting Math Concepts

Lesson 69

Part 1

a. 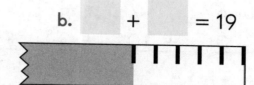 ▢ + ▢ = 61

b. ▢ + ▢ = 19

Part 2

a. 41 + 9 = _____

b. 68 + 9 = _____

c. 59 + 9 = _____

d. 77 + 9 = _____

e. 82 + 9 = _____

f. 16 + 9 = _____

Independent Work

Part 3 Write the sign >, <, or =.

a.   34 cents ▢ _____ cents

b. _____ cents ▢ 25 cents

c.   69 cents ▢ _____ cents

Lesson 69

Name _____

Part 4

| a. 5
 + 9 | b. 8
 − 3 | c. 6
 + 10 | d. 8
 + 3 | e. 9
 + 8 | f. 3
 + 7 | g. 2
 + 9 | h. 3
 + 8 | i. 9
 + 5 |

Part 5

a. 862
  − 336

b. 874
  − 337

Part 6

a. P < 15
    14 < P

_____

b. B > T
    T > Z

_____

c. J < 10
    10 < W

_____

Part 7

a. 10 + 25 = _____

b. 143 + 10 = _____

c. 250 + 10 = _____

d. 10 + 78 = _____

Copyright © The McGraw-Hill Companies, Inc.

Connecting Math Concepts

Lesson 70

Name _____

a. $72 + 9 =$ _____ b. $18 + 9 =$ _____ c. $44 + 9 =$ _____

d. $83 + 9 =$ _____ e. $51 + 9 =$ _____ f. $29 + 9 =$ _____

Independent Work

Part 2

a. $\begin{array}{r} 49 \\ + 53 \\ \hline \end{array}$ b. $\begin{array}{r} 62 \\ + 67 \\ \hline \end{array}$ c. $\begin{array}{r} 15 \\ + 85 \\ \hline \end{array}$ d. $\begin{array}{r} 94 \\ + 23 \\ \hline \end{array}$

Part 3

a. $53 +$ ☐ $=$ ☐   b. ☐ $+$ ☐ $= 81$

Part 4

a. $\begin{array}{r} 318 \\ + 43 \\ \hline \end{array}$ b. $\begin{array}{r} 466 \\ -418 \\ \hline \end{array}$ c. $\begin{array}{r} 789 \\ -653 \\ \hline \end{array}$ d. $\begin{array}{r} 87 \\ + 17 \\ \hline \end{array}$

Part 5

a. $11 > T$
$T > P$

_____

Part 6

a. $\begin{array}{r} 5 \\ + 9 \\ \hline \end{array}$ b. $\begin{array}{r} 11 \\ - 3 \\ \hline \end{array}$ c. $\begin{array}{r} 7 \\ + 9 \\ \hline \end{array}$ d. $\begin{array}{r} 16 \\ - 8 \\ \hline \end{array}$ e. $\begin{array}{r} 17 \\ - 6 \\ \hline \end{array}$ f. $\begin{array}{r} 18 \\ - 9 \\ \hline \end{array}$

g. $\begin{array}{r} 14 \\ - 7 \\ \hline \end{array}$ h. $\begin{array}{r} 4 \\ + 9 \\ \hline \end{array}$ i. $\begin{array}{r} 10 \\ - 3 \\ \hline \end{array}$ j. $\begin{array}{r} 12 \\ - 6 \\ \hline \end{array}$ k. $\begin{array}{r} 6 \\ + 9 \\ \hline \end{array}$ l. $\begin{array}{r} 10 \\ + 10 \\ \hline \end{array}$

Level C Correlation to Grade 2
Common Core State Standards for Mathematics

Operations and Algebraic Thinking (2.OA)

Represent and solve problems involving addition and subtraction.

1. Use addition and subtraction within 100 to solve one- and two-step word problems involving situations of adding to, taking from, putting together, taking apart, and comparing, with unknowns in all positions, e.g., by using drawings and equations with a symbol for the unknown number to represent the problem.

Lessons	WB 1: 33–41, 65 WB 2: 93, 94, 107, 121–125, 127, 128, 130 TB: 42–100, 102, 103, 105–112, 114–116, 118–120, 123, 129

Operations and Algebraic Thinking (2.OA)

Add and subtract within 20.

2. Fluently add and subtract within 20 using mental strategies. By end of Grade 2, know from memory all sums of two one-digit numbers.

Lessons	WB 1: 1–39, 41–70 WB 2: 71–80, 82, 85–122, 129, 130 TB:42–49, 56, 58, 59, 62–64, 68–73, 75, 77, 78, 81–83, 85–87, 95, 96, 99–106, 110–115, 119, 121–127

Operations and Algebraic Thinking (2.OA)

Work with equal groups of objects to gain foundations for multiplication.

3. Determine whether a group of objects (up to 20) has an odd or even number of members, e.g., by pairing objects or counting them by 2s; write an equation to express an even number as a sum of two equal addends.

Lessons	WB 1: 55–60 WB 2: 128, 129

Operations and Algebraic Thinking (2.OA)

Work with equal groups of objects to gain foundations for multiplication.

4. Use addition to find the total number of objects arranged in rectangular arrays with up to 5 rows and up to 5 columns; write an equation to express the total as a sum of equal addends.

Lessons	WB 1: 47, 48 WB 2: 71, 73 TB: 45–47, 49, 75

Number and Operations in Base Ten (2.NBT)

Understand place value.

1. Understand that the three digits of a three-digit number represent amounts of hundreds, tens, and ones; e.g., 706 equals 7 hundreds, 0 tens, and 6 ones. Understand the following as special cases:
a. 100 can be thought of as a bundle of ten tens — called a "hundred."
b. The numbers 100, 200, 300, 400, 500, 600, 700, 800, 900 refer to one, two, three, four, five, six, seven, eight, or nine hundreds (and 0 tens and 0 ones).

Lessons	WB 1: 1–4, 7, 10–25, 27, 29, 36, 59, 68 WB 2: 74, 116, 118, 126–129 Student Practice Software: Block 1 Activities 1 and 2, Block 4 Activities 2 and 3, Block 6 Activity 6

Number and Operations in Base Ten (2.NBT)

Understand place value.

2. Count within 1000; skip-count by *2, 5s, 10s, and 100s.

Lessons	WB 1: 11, 12, 15–29, 31–60, 63–65, 67–69 WB 2: 73–77, 80, 88, 90–94, 98, 101, 102, 116–120, 122–126, 130 TB: 41, 62, 63, 65, 68, 69, 72, 73, 76–80, 86, 87, 89 ,90, 95–117, 119, 120, 122, 125–127 Student Practice Software: Block 3 Activity 2

*Denotes California-only content.

Number and Operations in Base Ten (2.NBT)

Understand place value.

3. Read and write numbers to 1000 using base-ten numerals, number names, and expanded form.

Lessons	WB 1: 1–7, 11–37, 39, 40, 42, 43, 45–51, 54, 58, 59, 68 WB 2: 74 TB: 82, 118 Student Practice Software: Block 2 Activity 3, Block 4 Activity 4, Block 5 Activity 4

Number and Operations in Base Ten (2.NBT)

Understand place value.

4. Compare two three-digit numbers based on meanings of the hundreds, tens, and ones digits, using >, =, and < symbols to record the results of comparisons.

Lessons	WB 1: 31, 36 WB 2: 112, 113, 115, 119–121, 124–128 Student Practice Software: Block 2 Activity 5

Number and Operations in Base Ten (2.NBT)

Use place value understanding and properties of operations to add and subtract.

5. Fluently add and subtract within 100 using strategies based on place value, properties of operations, and/or the relationship between addition and subtraction.

Lessons	WB 1: 1–70 WB 2: 71–91, 93–130 TB: 47–50, 52–91, 94–96, 98–100, 102–115, 117–129 Student Practice Software: Block 3 Activities 1, 3, 4, 5; Block 5 Activity 2

Number and Operations in Base Ten (2.NBT)

Use place value understanding and properties of operations to add and subtract.

6. Add up to four two-digit numbers using strategies based on place value and properties of operations.

Lessons	WB 1: 4–10, 12, 16–20, 22–56, 60–70 WB 2: 71–94, 96–98, 100–102, 110, 111, 113, 122, 127–130 TB: 46, 47, 49, 50, 52–55, 57, 59–66, 69–71, 73, 75–82, 84, 86, 88, 89, 92–100, 102, 107–114, 117, 119, 122, 124, 125, 126, 128 Student Practice Software: Block 1 Activity 3

Number and Operations in Base Ten (2.NBT)

Use place value understanding and properties of operations to add and subtract.

7. Add and subtract within 1000, using concrete models or drawings and strategies based on place value, properties of operations, and/or the relationship between addition and subtraction; relate the strategy to a written method. Understand that in adding or subtracting three-digit numbers, one adds or subtracts hundreds and hundreds, tens and tens, ones and ones; and sometimes it is necessary to compose or decompose tens or hundreds.

Lessons	WB 1: 5–8, 10–70 WB 2: 71, 73–89, 92–95, 100, 106, 107, 109, 111–115, 119–121, 124, 125, 127–130 TB: 46–50, 62–69, 71, 73–83, 85, 86, 88–117, 119–123, 125, 127, 128 Student Practice Software: Block 1 Activity 5, Block 2 Activity 2, Block 5 Activity 5

Number and Operations in Base Ten (2.NBT)

Use place value understanding and properties of operations to add and subtract.

*7.1 Use estimation strategies to make reasonable estimates in problem solving.

Lessons	WB 2: 75–84, 88, 90–94, 97, 98, 100, 102 TB: 76–79, 95, 96, 98, 99, 105, 107–109, 111–114, 125, 126, 128

*Denotes California-only content.

Number and Operations in Base Ten (2.NBT)

Use place value understanding and properties of operations to add and subtract.

8. Mentally add 10 or 100 to a given number 100–900, and mentally subtract 10 or 100 from a given number 100–900.

Lessons	WB 1: 22–24, 42, 43, 50, 58, 59 WB 2: 75–78, 80, 89, 112, 117–121 TB: 89

Number and Operations in Base Ten (2.NBT)

Use place value understanding and properties of operations to add and subtract.

9. Explain why addition and subtraction strategies work, using place value and the properties of operations.

Lessons	WB 2: 115–120

Measurement and Data (2.MD)

Measure and estimate lengths in standard units.

1. Measure the length of an object by selecting and using appropriate tools such as rulers, yardsticks, meter sticks, and measuring tapes.

Lessons	WB 1: 30–41, 43, 44, 46, 49, 54, 59–68 WB 2: 115–120, 122, 124–127

Measurement and Data (2.MD)

Measure and estimate lengths in standard units.

2. Measure the length of an object twice, using length units of different lengths for the two measurements; describe how the two measurements relate to the size of the unit chosen.

Lessons	WB 1: 34, 35, 41, 43, 44, 46, 49, 51, 53, 54 WB 2: 85–87, 116

Measurement and Data (2.MD)

Measure and estimate lengths in standard units.

3. Estimate lengths using units of inches, feet, centimeters, and meters.

Lessons	WB 2: 85–87 TB: 125–127

Measurement and Data (2.MD)

Measure and estimate lengths in standard units.

4. Measure to determine how much longer one object is than another, expressing the length difference in terms of a standard length unit.

Lessons	WB 2: 115–120, 122

Measurement and Data (2.MD)

Relate addition and subtraction to length.

5. Use addition and subtraction within 100 to solve word problems involving lengths that are given in the same units, e.g., by using drawings (such as drawings of rulers) and equations with a symbol for the unknown number to represent the problem.

Lessons	TB: 52, 53, 57–71, 74, 75, 83, 84, 94, 96, 98, 123, 126

Measurement and Data (2.MD)

Relate addition and subtraction to length.

6. Represent whole numbers as lengths from 0 on a number line diagram with equally spaced points corresponding to the numbers 0, 1, 2, …, and represent whole-number sums and differences within 100 on a number line diagram.

Lessons	WB 1: 51–54, 56, 59, 67–70 WB 2: 125, 127, 128 Student Practice Software: Block 6 Activity 1

Measurement and Data (2.MD)

Work with time and money.

7. Tell and write time from analog and digital clocks to the nearest five minutes, using a.m. and p.m. ***Know relationships of time (e.g., minutes in an hour, days in a month, weeks in a year).**

Lessons	WB 2: 91–107, 109, 112, 114, 116, 118, 120, 122–124, 130 TB: 85–88, 90, 98, 99, 101, 103, 106, 108, 109, 112, 115, 117–119, 121–123, 126 Student Practice Software: Block 6 Activity 2

*Denotes California-only content.

Measurement and Data (2.MD)

Work with time and money

8. Solve word problems involving dollar bills, quarters, dimes, nickels, and pennies, using $ and ¢ symbols appropriately. *Example: If you have 2 dimes and 3 pennies, how many cents do you have?*

Lessons	WB 2: 94, 130 TB: 43, 46, 48, 54–56, 58, 62, 64, 65, 73, 75, 76, 80, 86, 87, 90, 93, 97–116, 118–121, 123, 124, 126, 128 Student Practice Software: Block 4 Activity 3, Block 5 Activity 6

Measurement and Data (2.MD)

Represent and interpret data.

9. Generate measurement data by measuring lengths of several objects to the nearest whole unit, or by making repeated measurements of the same object. Show the measurements by making a line plot, where the horizontal scale is marked off in whole-number units.

Lessons	WB 2: 124–127 Student Practice Software: Block 6 Activity 3

Measurement and Data (2.MD)

Represent and interpret data.

10. Draw a picture graph and a bar graph (with single-unit scale) to represent a data set with up to four categories. Solve simple put-together, take-apart, and compare problems using information presented in a bar graph.

Lessons	WB 2: 117–120, 122, 123, 125 TB: 121, 124, 126, 127 Student Practice Software: Block 6 Activity 4

Geometry (2.G)

Reason with shapes and their attributes.

1. Recognize and draw shapes having specified attributes, such as a given number of angles or a given number of equal faces. Identify triangles, quadrilaterals, pentagons, hexagons, and cubes.

Lessons	WB 1: 43–49, 54, 56–58 WB 2: 111–118, 127–129 TB: 50, 51, 53, 55–57, 59, 60, 66, 68, 70, 74, 76, 79, 84, 110 Student Practice Software: Block 4 Activity 5

Geometry (2.G)

Reason with shapes and their attributes.

2. Partition a rectangle into rows and columns of same-size squares and count to find the total number of them.

Lessons	WB 1: 47, 48 WB 2: 121–126, 128 TB: 45, 46, 48, 49, 126 Student Practice Software: Block 4 Activity 6

Geometry (2.G)

Reason with shapes and their attributes.

3. Partition circles and rectangles into two, three, or four equal shares, describe the shares using the words *halves, thirds, half of, a third of,* etc., and describe the whole as two halves, three thirds, four fourths. Recognize that equal shares of identical wholes need not have the same shape.

Lessons	WB 2: 125–127 Student Practice Software: Block 6 Activity 5

Standards for Mathematical Practice

Connecting Math Concepts addresses all of the Standards for Mathematical Practice throughout the program. What follows are examples of how individual standards are addressed in this level.

1. Make sense of problems and persevere in solving them.

Word Problems (Lessons 12–25, 28–77): Students learn to identify specific types of word problems (i.e., start-end, comparison, classification) and set up and solve the problems based on the specific problem types.

2. Reason abstractly and quantitatively.

Addition/Subtraction (Lessons 1–64, 73–115): Beginning in Lesson 45, students count objects in two groups, write the number for each group, and then add them to find the total objects. They connect the written numbers with quantities while learning the concept of addition.

3. Construct viable arguments and critique the reasoning of others.

Estimation (Lessons 65–94): Students learn how to round numbers and then apply that knowledge to word problems involving estimation. They work the original problem and the estimation problem and then compare answers to verify that the estimated answer is close to the exact answer. Students can construct an argument to persuade someone whether an estimated answer is reasonable.

4. Model with mathematics.

Number Families (Lessons 1–29): Students learn to represent three related numbers in a number family. Later, they apply number families to model and solve word problems.

5. Use appropriate tools strategically.

Throughout the program (Lessons 1–130) students use pencils, workbooks, lined paper, and textbooks to complete their work. They use rulers to measure lines. They use the computer to access the Practice Software where they apply the skills they learn in the lessons.

6. Attend to precision.

Measurement (Lessons 30–43, 59–78, 95–103): When measuring lines and finding perimeter and area, students learn to include the correct unit in the verbal and written answers. They also include units in answers to word problems that involve specific units.

7. Look for and make use of structure.

Mental Math (Lesson 9 and frequently throughout): Students build computational fluency by learning patterns inherent in different problem types, such as +/– 10 and +/– 100.

8. Look for and express regularity in repeated reasoning.

Multiplication (Lessons 32–49): Students are introduced to the concept of multiplication by thinking of it as repeated counting. For example, 2 x 5 tells us to count by 2 five times.